T0205625

Engineering Materials

This series provides topical information on innovative, structural and functional materials and composites with applications in optical, electrical, mechanical, civil, aeronautical, medical, bio- and nano-engineering. The individual volumes are complete, comprehensive monographs covering the structure, properties, manufacturing process and applications of these materials. This multidisciplinary series is devoted to professionals, students and all those interested in the latest developments in the Materials Science field, that look for a carefully selected collection of high quality review articles on their respective field of expertise.

Indexed at Compendex (2021) and Scopus (2022)

Sukanya Pradhan · Smita Mohanty
Editors

Bio-based Superabsorbents

Recent Trends, Types, Applications and Recycling

Editors
Sukanya Pradhan
Central Institute of Petrochemicals Engineering and Technology (CIPET)
SARP-Laboratory for Advanced Research in Polymeric Materials (LARPM)
Bhubaneswar, Odisha, India

Smita Mohanty
Central Institute of Petrochemicals Engineering and Technology (CIPET)
SARP-Laboratory for Advanced Research in Polymeric Materials (LARPM)
Bhubaneswar, Odisha, India

ISSN 1612-1317 ISSN 1868-1212 (electronic)
Engineering Materials
ISBN 978-981-99-3096-8 ISBN 978-981-99-3094-4 (eBook)
https://doi.org/10.1007/978-981-99-3094-4

This Springer imprint is published by the registered company Springer Nature Singapore Pte Ltd.
The registered company address is: 152 Beach Road, #21-01/04 Gateway East, Singapore 189721, Singapore

Preface

Superabsorbent polymers have created a very attractive area in the viewpoint of super-swelling behaviour, chemistry and designing a variety of final applications. A bio-based resource would be the perfect choice to replace fossil-based resources from a sustainability standpoint. Bio-based and recyclable superabsorbent polymers are necessary to address the inadequacies of conventional superabsorbents. Both the industrial and academic research and the commercial relevance of superabsorbent polymers have grown over the years. Hence, this book will have a broad audience which can serve as a reference book at both academic and industrial levels. It can also serve as a textbook for students in polymer science and engineering, plastics technology, materials science, etc. This will be an excellent book for scientists, engineers, graduate students and industrial researchers in the field of bio-based materials.

However, there are no recent books available in the market on this recent topic. This book systematically describes the green engineering, chemistry and synthesis of bio-based superabsorbent polymers and composites. This book will stand apart with its unique combination of all the basic principles, classification and special focus on bio-based superabsorbent along with its recycling and reuse for various applications.

In the chapters, the book defines the importance of natural superabsorbent polymers over synthetic-based counterparts. It will describe various novel approaches for the fabrication of absorbent polymers with advanced properties. The increasing literature on SAP research published recently will be examined in this book. The objective of the book will also include the issues behind the sharp increase in research attention, namely what the prevailing research hotspots and clusters are and what preferences they suggest with regard to present studies, what works have been significant and pivotal in the development of SAP research, and what are the current advances and future directions of research. It will also examine the emerging applications of superabsorbent polymers.

Bhubaneswar, India

Sukanya Pradhan
Smita Mohanty

Contents

Bio-based Superabsorbent Polymers: An Overview 1
Jitender Dhiman, Kumar Anupam, Vaneet Kumar, and Saruchi

Synthesis of Biopolymer Based Superabsorbent: An Eco-friendly Approach Towards Future Sustainability 29
Sweta Sinha

Bio-based Versus Petro-based Superabsorbent Polymers 51
Shiv Kumari Panda

Theory of Superabsorbent Polymers 67
Sulena Pradhan and Sukanya Pradhan

Chitin/Chitosan Based Superabsorbent Polymers 77
Swarnalata Sahoo

Alginate-Based Superabsorbents 93
D. Thirumoolan, T. Siva, R. Ananthakumar, and K. S. Nathiga Nambi

Starch-Based Superabsorbent Polymer 115
Jaylalita Jyotish, Rozalin Nayak, Debajani Tripathy, Srikanta Moharana, and R. N. Mahaling

Smart Superabsorbents and Other Bio-based Superabsorbents 145
Shubhasmita Rout

Recycling and Reuse of Superabsorbent Polymers 161
Ankita Subhrasmita Gadtya, Debajani Tripathy, and Srikanta Moharana

Bio-based Superabsorbent Polymer: Current Trends, Applications and Future Scope 185
Roshni Pattanayak and Tapaswini Jena

Correction to: Smart Superabsorbents and Other Bio-based Superabsorbents C1
Shubhasmita Rout

Editors and Contributors

About the Editors

Sukanya Pradhan is currently working as a Research Scientist at LARPM—Central Institute of Petrochemicals Engineering & Technology (CIPET), Ministry of Chemicals and Fertilizers, Department of Chemicals and Petrochemicals, Govt. of India. She has received her Ph.D. in polymers and plastics technology from Anna University, India, in 2019. She has published over 20 papers in peer-reviewed journals, book chapters and international conference proceedings. Her research interest includes Polymer synthesis, Polymer characterization coatings and adhesive, Polymer composites, Biodegradable polymers and feminine hygiene. She has presented many invited and contributed talks at international conferences. She has over 5 years of teaching experience at undergraduate, graduate and post-graduate levels. She has active participation in applied innovative teaching methods to encourage student learning objectives. She is also a life member of various professional bodies (Indian Science Congress Association, Orissa Chemical Society, etc.).

Smita Mohanty is the Principal Scientist at School for Advanced Research in Petrochemicals (SARP): LARPM—Central Institute of Petrochemicals Engineering & Technology (CIPET), Ministry of Chemicals and Fertilizers, Department of Chemicals and Petrochemicals, Government of India. She has over 15 years of experience, with research interests in polymers, coatings, adhesives, paints and composites, biomaterials, and functionalization of polymeric materials. She has published over 160 articles in international peer-reviewed journals and is the winner of the National Awards by the Department of Chemicals and Petrochemicals, Govt. of India, and Young Scientist Award by Department of Science and Technology (DST), Govt. of India. She was a recipient of Outstanding Research Faculty 2017–18 in the Materials Science discipline through Scopus by Careers 360 and has 7 Indian patents to her credit.

Contributors

R. Ananthakumar Laboratory for Advanced Research in Polymeric Materials (LARPM), School for Advanced Research in Petrochemicals, Central Institute of Petrochemicals Engineering and Technology (CIPET), Bhubaneswar, India;
Advanced Research School for Technology and Product Simulation (ARSTPS), School for Advanced Research in Polymers (SARP), Central Institute of Petrochemical Engineering & Technology (CIPET), Tamil nadu, Chennai, India

Kumar Anupam Chemical Recovery and Biorefinery Division, Central Pulp and Paper Research Institute, Saharanpur, Uttar Pradesh, India

Jitender Dhiman Biotechnology Division, Central Pulp, and Paper Research Institute, Saharanpur, Uttar Pradesh, India

Ankita Subhrasmita Gadtya School of Applied Sciences, Centurion University of Technology and Management, R.Sitapur, Odisha, India

Tapaswini Jena CIPET: SARP-LARPM, Bhubaneswar, India

Jaylalita Jyotish Laboratory of Polymeric and Materials Chemistry, School of Chemistry, Sambalpur University, Jyoti Vihar, Burla, Odisha, India

Vaneet Kumar Department of Applied Sciences, CTIEMT, CT Group of Institutions, Jalandhar, Punjab, India

R. N. Mahaling Laboratory of Polymeric and Materials Chemistry, School of Chemistry, Sambalpur University, Jyoti Vihar, Burla, Odisha, India

Srikanta Moharana School of Applied Sciences, Centurion University of Technology and Management, R.Sitapur, Paralakhemundi, Odisha, India

K. S. Nathiga Nambi Department of Biology, The Gandhigram Rural Institute (Deemed to Be University), Gandhigram, Dindigul District, Tamil Nadu, India

Rozalin Nayak Laboratory of Polymeric and Materials Chemistry, School of Chemistry, Sambalpur University, Jyoti Vihar, Burla, Odisha, India

Shiv Kumari Panda Department of Chemistry, U.N. Autonomous College of Science & Technology, Adaspur, Cuttack, Odisha, India

Roshni Pattanayak CIPET: SARP-LARPM, Bhubaneswar, India

Sukanya Pradhan SARP-LARPM, CIPET, Bhubaneswar, Odisha, India

Sulena Pradhan Larodan AB, Karolinska Institute, Solna, Sweden

Shubhasmita Rout CIPET-IPT, Bhubaneswar, Odisha, India

Swarnalata Sahoo CIPET: IPT, BBSR, Bhubaneswar, Odisha, India

Saruchi Department of Biotechnology, CT Group of Institutions, Jalandhar, Punjab, India

Sweta Sinha Department of Chemistry, Amity Institute of Applied Sciences, Amity University Jharkhand, Ranchi, Jharkhand, India

T. Siva Laboratory for Advanced Research in Polymeric Materials (LARPM), School for Advanced Research in Petrochemicals, Central Institute of Petrochemicals Engineering and Technology (CIPET), Bhubaneswar, India;
Advanced Research School for Technology and Product Simulation (ARSTPS), School for Advanced Research in Polymers (SARP), Central Institute of Petrochemical Engineering & Technology (CIPET), Tamil nadu, Chennai, India

D. Thirumoolan Department of Chemistry, Annai College of Arts and Science (Affiliated to Bharathidasan University), Kovilacheri, Kumbakonam, Tamil Nadu, India

Debajani Tripathy School of Applied Sciences, Centurion University of Technology and Management, R.Sitapur, Paralakhemundi, Odisha, India

Bio-based Superabsorbent Polymers: An Overview

Jitender Dhiman, Kumar Anupam, Vaneet Kumar, and Saruchi

Abstract Superabsorbents (SAs) are magic materials that became a point of interest in the research community over a long period. Several studies related to these materials are conducted to manufacture diverse types of superabsorbents to meet the needs of the industrial sector. Superabsorbents are materials capable of absorbing water or fluids about 1000 times in comparison to their weight in dry form and release them in a consistent and slow manner. Due to this property of huge water/fluids absorption, these materials have great applications in the production of hygiene products and the medical field. In 2019, only in the hygiene market, SAs contributes about USD 9.58 billion. SAs are synthetic and generate huge environmental issues; as a consequence of this, bio-based SAs draw the attention of the research community due to their biodegradable, renewable, and biocompatible nature. This chapter focuses on the classification, methods of synthesis, and characterization of SAs. It also provides an overview of their applications in different fields including agriculture, wastewater treatment, agriculture, biomedical, and hygiene. This Chapter also performed an analysis of research trends in SAs during the last decade (2011–2021) to visualize major research progress in this field. The task is served by title analysis, abstract analysis, keyword analysis, and publication year's analysis in the number of publications on SAs.

Keywords Superabsorbents · Water absorption · Hygiene market · Medical field · Wastewater treatment · Swelling · Bio-based

---

J. Dhiman (✉)
Biotechnology Division, Central Pulp, and Paper Research Institute, Saharanpur, Uttar Pradesh, India
e-mail: jitenderdhiman81@gmail.com
URL: https://scholar.google.com/citations?user=hQtperEAAAAJ&hl=en

K. Anupam
Chemical Recovery and Biorefinery Division, Central Pulp and Paper Research Institute, Saharanpur, Uttar Pradesh, India

V. Kumar
Department of Applied Sciences, CTIEMT, CT Group of Institutions, Jalandhar, Punjab, India

Saruchi
Department of Biotechnology, CT Group of Institutions, Jalandhar, Punjab, India

S. Pradhan and S. Mohanty (eds.), *Bio-based Superabsorbents*, Engineering Materials,
https://doi.org/10.1007/978-981-99-3094-4_1

1 Introduction

Superabsorbents (SAs) are smart materials that can absorb a significant amount of fluid than its dry weight. Upon swelling due to the up-take of a huge amount of fluid, these can still maintain their structural properties. SAs possesses three-dimensional crosslinked structure which is porous in nature and highly swellable on coming in contact with fluids such as water, blood, urine, electrolyte solution, etc. One gram of SAs can hold fluid from 200 g to 4 kg, depending on its nature, without getting broken down. Although, Superabsorbents are somewhat dissimilar to general hydrogels in regards of fluid uptake capacity, larger stretchability in dry form, crosslinking and porosity. It is a tedious task to eliminate the fluid from these types of materials in comparison to other fluid-absorbing materials such as cotton, sponge, napkins, etc. [51, 71].

Initially, the swellable polymers were prepared using acrylic acid and divinylbenzene. Later, SAs were started to be prepared from hydroxyalkyl methacrylate and other related monomers. These were stated to use as contact lenses from 1960. These kinds of polymers have a 40–50% swelling capacity and were known as first-generation polymers. The second generation of these polymers was able to swell 70–80% which increase their scope across various fields. Firstly, Japan utilized these SAs in the hygiene industry and commercialized their use. Later, Germany and France utilized these polymers in the production of baby diapers; this step spikes the production of SAs up to 32,000 metric tons yearly. Simultaneously, some countries started to use these polymers in making napkins which absorbs fluid quickly and retain fluid under pressure, therefore, it became an ideal material for making sanitary pads and diapers [8, 54, 101, 115].

SAs are generally categorized into two types: one of them is petroleum-based (synthetic superabsorbents) and the other is polysaccharide and polypeptide-based (natural superabsorbents). Generally, monomers such as acrylamide (AM) and acrylic acid (AA) and its salts are utilized for the manufacturing of SAs with the help of polymerization technique. A huge range of SAs products is conveniently available in market due to the reason of its variability according to its application and cost-effectiveness. These materials can be easily bought in the range of US$10 to $40 per kilogram. SAs have a wide range of applications across many disciplines such as agriculture, effluent treatment, medical field, textile industry, civil engineering, soil science, and in hygiene industry. Some of the special categories of SAs can respond to specific molecules; thus acts as a biosensor and drug delivery equipment [10, 61, 86, 110, 116].

In paper industry, the SAs are used for fluid absorbents in paper diapers. Different types of diapers are available in the market according to costumer needs. Generally, diapers are categorized into two types: one is baby diapers (baby disposable diaper, baby training diaper, baby cloth diaper, baby swim pants, and others); the second type is adult diapers (adult pad type diaper, adult flat type diaper, adult pant type diaper). Paper diapers consist of mainly three components, i.e., pulp (exterior material), superabsorbents (fluid absorbent), and plastic (internal fabric material). Pulp is extracted from various plants such as bamboo, hemp, cotton, etc., and plastic such as polyethylene and polypropylene is utilized for the inner lining of diapers. In India, the diaper industry touches US$ 5.93 billion in 2022. This industry is projected to grow by 7.58% annually during 2022–2027. By 2027, the volume of baby diaper market segment is anticipated to be 1324.4 mKg. The consumption of baby diapers per individual in terms of volume is 0.8 kg in 2022 [38].

Therefore, from the above discussion, the present chapter focuses on the classification, synthesis methods, and characterization of SAs. This chapter also focuses on different applications of SAs in multiple fields such as agriculture, wastewater treatment, biomedical, and hygiene. Also, we performed an analysis of research trends in SAs during the last decade (2011–2021) via visualize scientific progress in this field. The analysis is done by analyzing the titles, keywords, publication years, and types of publications on SAs.

2 Classification of Bio-based Superabsorbents

Classification of bio-based superabsorbents can be performed based on source or origin, crosslinking, electrical charges, morphology, water affinity, respond to stimuli, and polymerization mechanism/methods (Fig. 1).

2.1 *Based on Source*

On the origin basis SAs are generally divided in two categories polysaccharides based (Fig. 2) and protein-based (Fig. 3). Different polysaccharides such as alginic acid, cellulose, chitosan, gellan gum, and hyaluronic acid can act as building blocks

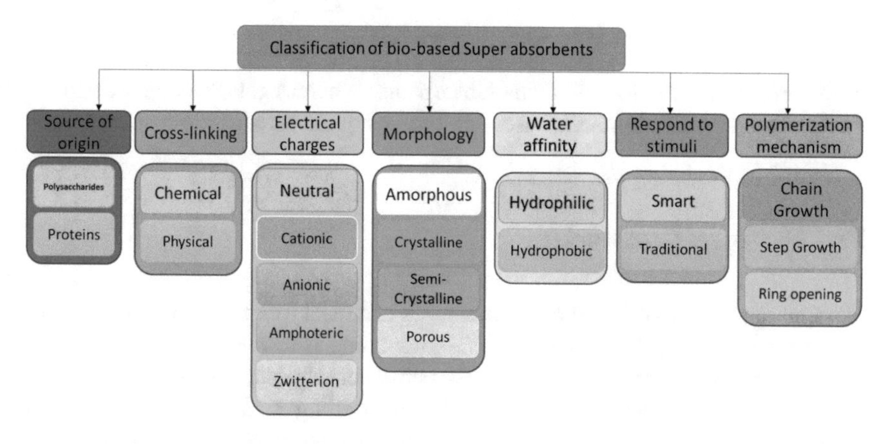

Fig. 1 Classification of bio-based superabsorbents

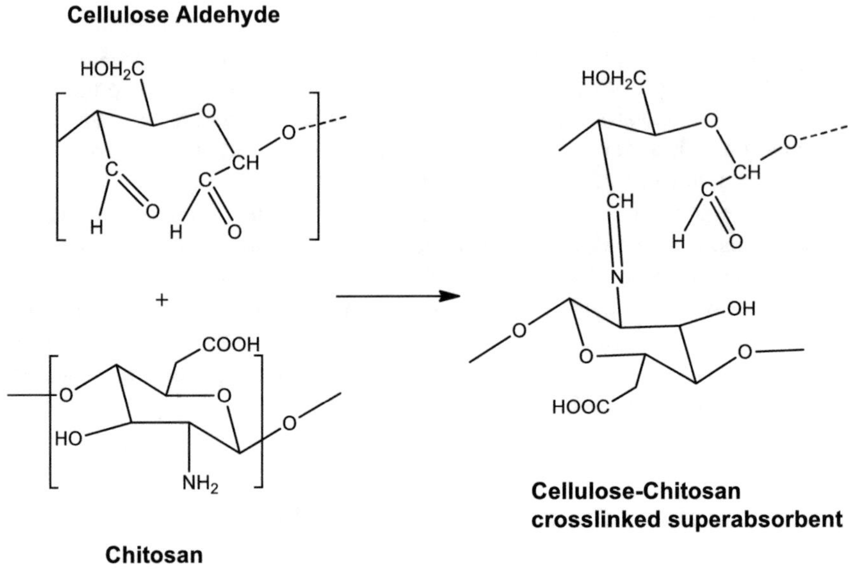

Fig. 2 Polysaccharide-based superabsorbent [4]

Fig. 3 Protein-based superabsorbent [12]

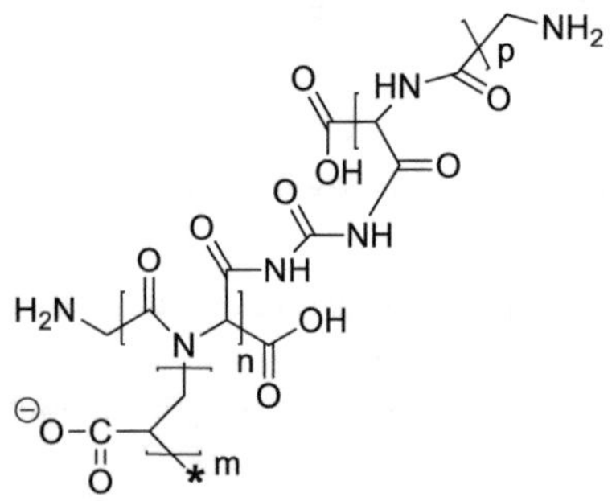

for the preparation of bio-based superabsorbents. Whereas gelatin/collagen and silk fibroin are examples of protein-based bio-based superabsorbents.

2.2 *Based on Crosslinking*

Crosslinking bases SAs are of two types one is chemical (permanent) SAs and physical (reversible) SAs. In chemical SAs, the bonding is covalent bonding which is permanent in nature and responsible for the permanent structure of chemical superabsorbents. These absorbents reach an equilibrium swelling condition that is dependent on the crosslink density and the polymer–water interaction parameter. On the other hand, in physical SAs secondary interactions, such as ionic or hydrogen bonding are responsible for their structural integrity. Since these interactions are reversible in nature, therefore, these SAs can change based on the application of stress or change in physical parameters [68, 82, 83, 107].

2.3 Based on Network Electrical Charges

Categorization of SAs can be performed based on the electrical charge they possess. Generally, SAs are neutral, cationic (Fig. 4), anionic (Fig. 5), amphoteric (Fig. 6), and zwitterionic (Fig. 7). SAs possessing no charge are termed neutral SAs as these are non-ionic in nature. Polyacrylamide grafted hydroxyethyl cellulose is an example of neutral SAs. Hydroxymethyl cellulose-based is another example of the same type of SAs. Ionic SAs are both cationic and anionic in nature. Acrylic monomers or sodium/potassium-based absorbents are the example of anionic SAs, whereas chitosan and acrylic monomer or *N*, *N*-diallyl, *N*, *N*-dimethyl ammonium chloride (DADMAC) with *N*-vinyl 2-pyrrolidone (NVP) are the forms Cationic SAs. Amphoteric SAs are those which possess both cationic and anionic functional groups, e.g., SA synthesized using maleic anhydride or N-vinyl succinimide can be amphoteric in nature [45]. Zwitterionic superabsorbents possess both positive and negative charges, but they neutralize charge of each other; resulting in a net zero charge, therefore, researchers synthesized such SA using poly(3-acrylamidopropyl) trimethylammonium chloride and 2-acrylamido-2-methylpropanesulphonic acid by free radical polymerization [79].

2.4 Based on Morphology

The morphologic classification of SAs consists of amorphous, semicrystalline, crystalline, and porous superabsorbents. Hydrogel prepared from N,N-dimethylacrylamide or polyacrylic acid chains and copolymerized using (meth)acrylate is an example of semicrystalline SA [46], which has self-healing properties. In another case, researchers used N, N^1-dimethylacrylamide and n-octadecyl acrylate to serve the same purpose; this SA possesses properties such as shape memory and self-healing [69]. SAs are composed using acrylic acid monomer are reported to be amorphous in nature. Carboxymethylcellulose and silver nanoparticles were in order to produce amorphous SAs [21]. These types of SAs have great contributions in the medical field specifically in bond healing. Crystalline SAs were prepared using acrylic acid and 11-(4′-cyanobiphenyloxy) undecyl acrylate to study the ordered structure and its formation in biological tissues [63]. Micro-porous SAs possesses properties that can efficiently deliver necessary nutrients and oxygen to desired cells. Researchers had manufactured these types of SAs using gelatine methacrylate and ethylene glycol [63].

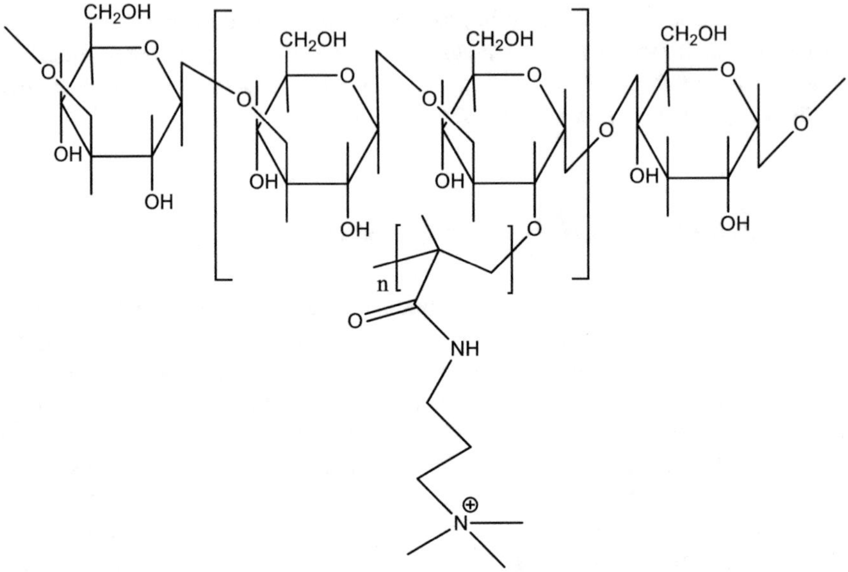

Fig. 4 Cationic superabsorbent [105]

2.5 Based on Water Affinity

It is possible to refer to SAPs as hydrophilic or hydrophobic depending on their affinity for water or oil. The former type is water loving and oil hating known as hydrophilic, whereas the latter is oil-loving and water-hating named as hydrophobic. Hydrophilic SAs are used in agriculture, medical field, food encapsulation, hygiene products, drug delivery, crack repair, bio-sensing, and wastewater treatment [13, 75, 78]. Hydrophobic SAs are utilized for oil/water separation, rust prevention, surgical tools, anti-icing, prevention of marine organisms from growing on a ship's hull, medical equipment, and oil spill cleaning [48, 53, 104].

2.6 Based on Respond to Stimuli

Smart SAs are the superabsorbents that change their three-dimensional conformation in response to some stimuli; without which the SAs are referred to be traditional or non-responsive to any of the environmental stimuli (Fig. 8). This type of SAs can be influenced by multiple stimuli such as temperature, pressure, electric/ magnetic field, pH, and ionic strength. The distortion is non-permanent in nature and as soon as the influence of stimuli is removed, these SAs regain their original state. Smart SAs are utilized in different applications such as microfluidic, drug delivery, diabetic/chronic wound dressing, biomaterial, and water remediation applications [30, 52, 60, 70, 91].

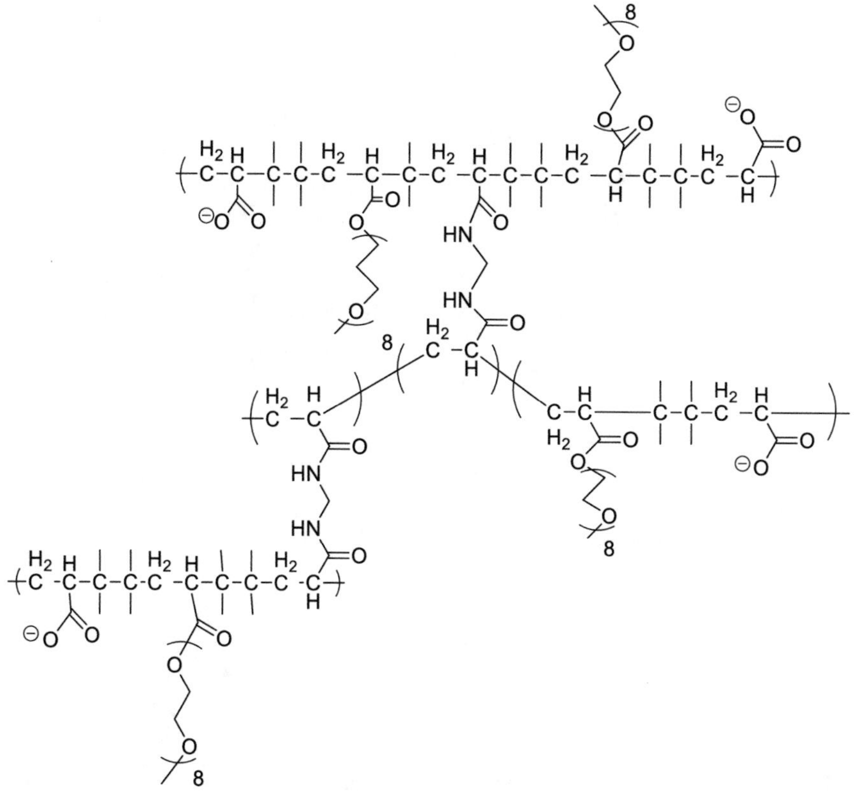

Fig. 5 Cationic superabsorbent [102, 103]

Fig. 6 Amphoteric superabsorbent [111]

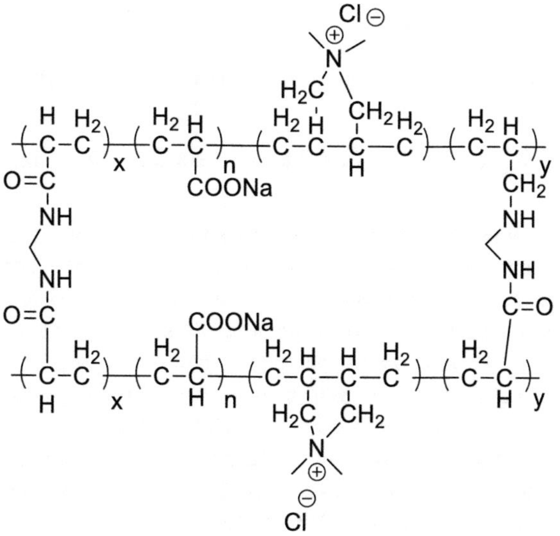

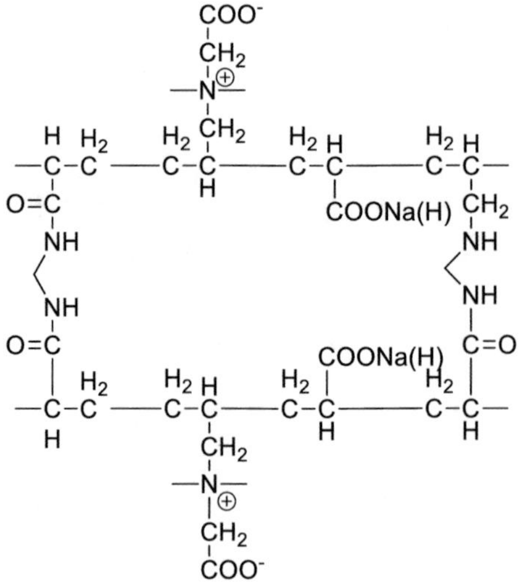

Fig. 7 Zwitterionic superabsorbent [106]

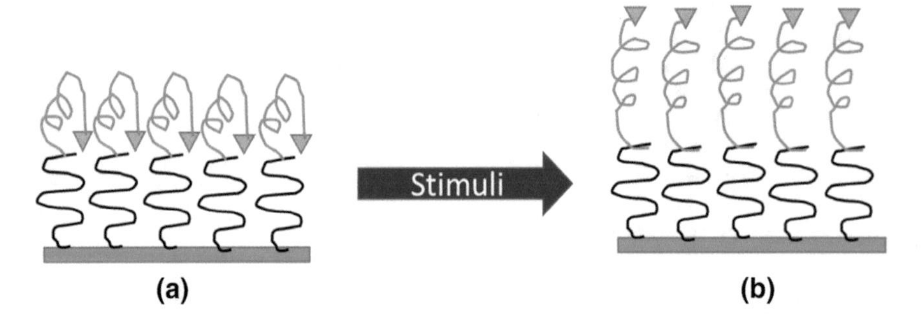

Fig. 8 **a** Functional group hidden. **b** Functional group exposed [77]

2.7 Based on Polymerization Mechanism/Methods

Mainly three types of polymerization mechanisms are commonly used for the synthesis of SAs, which include chain growth, step growth, and ring-opening mechanism. Chain growth mechanism incudes the free radical of polymerization. Synthesis is generally initiated by the generation of free radicals which originated from initiators like ammonium persulfate, potassium persulfate or mixture of both used in this process. The reaction proceeds with the polymer growth step in which the poly(monomer) chains are attached to the backbone. Finally, the process stopped with the suppression of free radicals in the termination step [18, 27, 89, 108, 112]. The step growth mechanism is also referred to as condensation polymerization. This mechanism executes with the formation of a long chain by two or more monomers along with the elimination of simple molecules like water, ammonia, etc. Step-growth polymerization involves interactions between active sites typically functional groups like alcohol, acid, and isocyanate. To create a lengthy polymer chain, the monomers must have at least two functional groups on them. A chain becomes branched when the functionality is greater than two, which can eventually cause crosslinking and create a thermoset polymer [55, 72, 102, 103].

Ring-opening polymerization (ROP) is a type of chain-growth polymerization which when chemically reactive center like cation, anion or free radical attacks on a cyclic monomer resulting in the formation of the longer polymer. In some cases, metal catalysis is also used for the ring-opening polymerization. When traditional chain-growth polymerization of vinyl monomers is not an option, radical ROP can be used to create polymers with functional groups integrated into the backbone chain. For instance, radical ROP can create polymers containing functional groups such as ethers, esters, amides, and carbonates along the main chain [19, 35, 47, 96, 97].

3 Synthesis methods of Superabsorbents

Superabsorbents are designed using a variety of synthetic procedures. Mainly, the synthesis methods can be split up into two groups: chemical synthesis and physical synthesis. Bulk polymerization, crosslinking, suspension polymerization, and radiation-influenced polymerization are all different forms of chemical synthesis methodologies. Furthermore, physical techniques use crosslinked hydrogen bonds and freeze–thaw cycle technology. Despite these two, some new technologies like interface contact technology and in-situ polymerization are also being used in recent days to serve the above-written purpose Fig. 9.

3.1 Chemical Synthesis

3.1.1 Bulk Polymerization

The bulk polymerization process consists of two phases. Initially, monomer-soluble initiators generate free radicals that polymerized vinyl chloride monomer up to 10%. In the later phase, more monomer is added which is responsible for further polymerization up to 80–85%. By the action of heat and vacuum, unreacted monomer portion is carried away. Bulk polymerization is a cost-effective method as in this process there is no obligation for drying during the polymerization process. The simplicity, consistency of resin particle size, high porosity of resin particles, and purity of the polymer are all benefits of the bulk process. The key weakness of this technique is the lack of product mix flexibility (homopolymers only). This technique is easiest to articulate since it can operate without the use of a solvent or dispersion [9, 11, 81, 87]. A pH-responsive, superabsorbent hydrogel was synthesized by bulk polymerization of sodium acrylate and hydroxyethyl methacrylate [92].

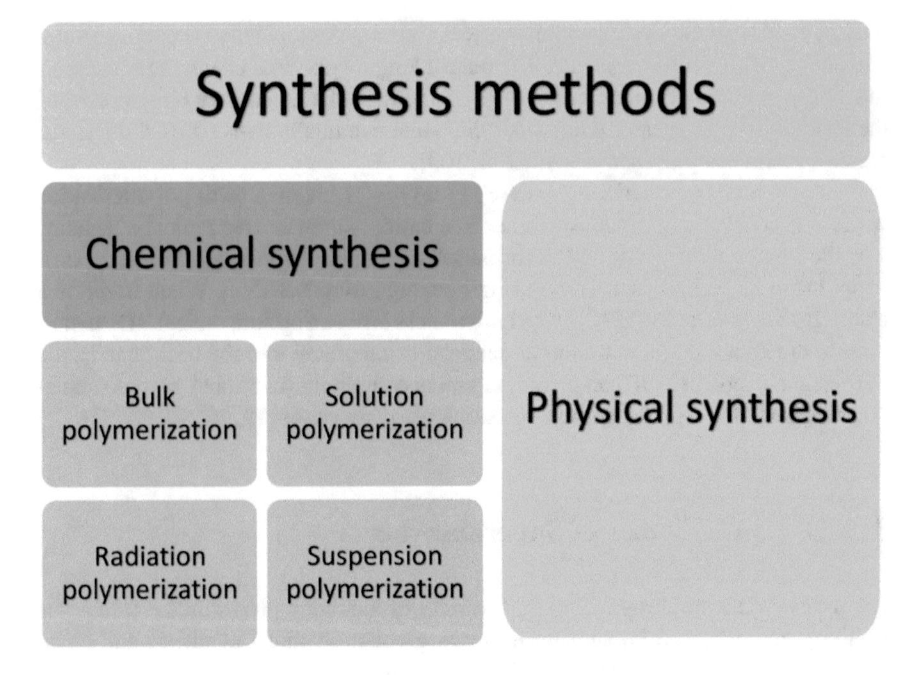

Fig. 9 Synthesis methods of superabsorbents

3.1.2 Solution Polymerization

The solvent is used in the reactor during solution polymerization, and it acts as a heat sink. Water, benzyl alcohol, ethanol as well as combinations of typical solvents are employed in the solution polymerization process. Solvent and initiator are poured in the initial stage of the procedure. Solvent selection is a very crucial step as the wrong solvent may restrict/terminate the polymer growth process. Solvent or combination of solvents is selected in such a way that all monomers, initiators, and final polymer should be soluble in it. The first initiator generates chemically unstable species commonly free radicals which initiate the process of polymerization. These free radicals keep on interacting with monomers to generate monomer free radicals that keep on adding till the end of the polymerization process. The decomposition of the initiator is a crucial factor that impacts the polymerization process, i.e., the efficiency of the initiator plays a vital role in the process. Since some of the initiator molecules react with undesired molecules rather than monomer molecules or self-terminated, therefore, the efficiency of the initiator is always less than 100%. A versatile crosslinking agent is blended with neutral or ionic monomers and thermal polymerization is initiated by UV-irradiation or a redox initiator mechanism. After the synthesis of desired superabsorbent, the solvent may be removed during a subsequent stage of hydration in water [1, 41, 57, 65].

3.1.3 Suspension Polymerization

One of the most often used techniques for producing SAs is suspension polymerization. This technique produces spherical SAP microparticles with a size range of 1 μm to 1 mm. The reaction occurs in a heterogeneous medium in which there is a mixture of a liquid matrix and droplets of monomer. The monomer solution is disseminated in the non-solvent during suspension polymerization, generating tiny monomer droplets that are stabilized by the addition of a stabilizer. The polymerization process starts with initiator decomposition under the influence of heat. Initially, an organic solvent is used to dissolve the functional monomer, crosslinker, and initiator. Then dispersant is then turned into solution form via dissolving it in water. Further to create a suspension, the organic and water phases are stirred together. Finally, the polymerization leads to the formation of homogeneous particles. By varying the stirring speed and volume of the organic and aqueous phases, the particle size of imprinted polymer microspheres can be regulated. The unreacted monomers, initiator, and crosslinking are subsequently washed off to ensure contamination-free microparticles. The microparticles generated possesses an average diameter that is comparable to the initial monomer droplet diameter (0.01–1 mm). Filtration can easily isolate the resulting polymer beads if they are stiff and not sticky. Thus, suspension polymerization is inappropriate for producing polymers with low glass transition temperatures, yet it is frequently employed to produce monomers such as methyl methacrylate, styrene, hydroxy ethyl methacrylate, vinyl chloride, etc. [29, 42, 64, 86, 100].

3.1.4 Radiations Polymerization

Polymerization of liquid ethylene was the first application of the irradiation polymerization method. The radiation, containing alpha rays, beta rays, gamma rays, and electron beams had been applied as a reaction initiator to create the SAs. Depending on the experimental setting and the monomer, which is under consideration, high-energy radiation beams may start either free radical or ionic polymerizations. Free radicals generated due to action of radiation beams bombardment on monomers results in the formation poly(monomer) radical chains that interact with backbone and generate desired polymer. Additionally, in the presence of a solvent, the radiolysis of solvent molecules and monomer generates free radicals, which attack polymer chains and create macro radicals as a result. Covalent bonds are generated due to the consequences of the macro radicals recombining on various chains, and as a result, a crosslinked structure is finally generated without the presence of any catalyst or initiator residue in the system. A very low temperature is essential to conduct this reaction. Cationic polymerizations occur in special conditions of low temperatures and with "ultra-dry" monomers. Due to the extremely high propagation rates of free ion propagation, enormously lengthy kinetic chains develop in these systems. Polymerization in solid-state take place in monomers having crystalline form at temperature just higher than their melting points. This effect seems to be the result of severe crystal dehydration that favors a cationic mechanism [6, 14, 36].

3.2 Physical Synthesis

Physical synthesis techniques involve assembly at the molecular level through crosslinking the ionic or hydrogen bonds between polymers. In comparison to procedures utilized at ambient temperatures, the superabsorbents can be made at lower temperatures using physical synthesis. The polymerization using physical synthesis occurs by virtue of strong hydrogen bonding. These strong hydrogen bonds come into play during the freeze or thaw cycle. This bonding may occur during the storage of the sample in cold conditions or during the thawing of the chilled sample. Repeated freeze–thaw cycles cause the creation of microcrystals, which serve as crosslinking sites and result in the production of hydrogel networks [31, 33, 44, 67, 95]. Researchers prepared SAs using vinyl alcohol monomer with the help of freeze/thaw cycles [67].

4 Characterization of Superabsorbents

A number of bio-based superabsorbents are prepared in the past by the scientific community all over the world (Table 1). They synthesize superabsorbents using a bio-waste material (Backbone), monomer, initiator, crosslinker, and solvent. These SAs further characterized using different characterization techniques such as FTIR, SEM, XRD, NMR, TEM, BET, TGA, EDS, and DSE (Fig. 10).

5 Applications of Bio-based Superabsorbents

Superabsorbents made of biomaterials have numerous uses in a variety of industries, including biomedicine, agriculture, construction, and water recycling (Fig. 11).

5.1 Biomedical Applications

Superabsorbents have characteristics like those of real tissues because of their properties including high biocompatibility, huge fluid uptake capacity, and water-loving qualities, making them useful for a variety of biomedical applications. It has been reported that these materials have applications in tissue engineering, anti-microbial materials, wound bandaging, and medication delivery. Contact lenses were the first application for superabsorbents in the biomedical field. Soft contact lenses were synthesized by polymerizing hydroxyethyl methacrylate in the 1950s which were further patented in 1960. For the continuous and controlled release of medication over a long period of time in the gut, superabsorbents are being utilized. They are

Table 1 Techniques used for characterization of superabsorbents

Sr. No.	Name of superabsorbent	Monomers used	Initiator	Characterization technique	References
1	Oil palm empty fruit bunch graft (acrylic acid-co-Ac-rylamide) superabsorbent composite	Acrylic acid, acrylamide	Ammonium persulphate	TGA, NMR, XRD	[116]
2	Starch polyacrylamide graft co-polymer	Acrylamide	Ceric ammonium nitrate	FTIR	[49]
3	Acrylonitrile and acrylic acid grafted on chitosan	Acrylonitrile, acrylic acid	Ammonium persulphate	FTIR	[94]
4	Itaconic acid grafted starch	Itaconic acid	Potassium permanganate, sodium bisulfite	FTIR, XRD and TEM	[24]
5	Dextrin-graft-acrylic acid/ Montmorillonite	Acrylic acid	Ammonium persulfate, sodium sulfite	–	[73]
6	Cassava starch-g-acrylamide/ itaconic acid	Acrylamide, itaconic acid	Ammonium persulphate, tetramethyl-ethylene diamine	FTIR, XPS	[43]
7	Chicken feathers proteing-poly (potassium acrylate)/polyvinyl alcohol	Acrylic acid, vinyl alcohol	Potassium persulfate, ammonium cerium nitrate	SEM, FESEM	[34]
8	SAP microspheres	Acrylamide, itaconic acid	Potassium persulfate	Fluorescence microscopy	[109]
9	Poly(acrylic acid) based nanocomposite hydrogel	Acrylic acid	Ammonium persulfate	TEM, SEM, EDS, FTIR, XRD	[15]
10	Superabsorbent polymer	Acrylic acid, acrylamide	Potassium persulfate	FTIR	[32]
11	Carboxymethyl cellulose-acrylic acid-superabsorbent polymer	Acrylic acid, vinyl alcohol	Benzoyl peroxide	FRIT, SEM	[39]
12	PAA-g-CMPWS	Acrylic acid	Ammonium persulfate	FTIR, SEM, TGA	[50]
13	AA and AM copolymerized hemicellulose	Acrylic acid (AA), acrylic amide	Ammonium persulfate, sodium hydrogen sulfite	FTIR, SEM, AFM	[113]
14	Collagen-g-poly(AAco-IA)	Acrylic acid, itaconic acid	Ammonium persulfate	FTIR, SEM, TGA	[93]
15	MB/Aam	Acrylamide	Ammonium persulfate	SEM, TGA, FTIR	[5]

(continued)

Table 1 (continued)

Sr. No.	Name of superabsorbent	Monomers used	Initiator	Characterization technique	References
16	Sugarcane bagasse-g-poly(acrylic acid-co-acrylamide) (SB/ P(AA-co-AM)) hydrogels	Acrylic acid, acrylamide	Ammonium persulfate, sodium sulfite	SEM, XRD, FTIR, DSC	[80]
17	CTS-AA-AN	Acrylic acid, acrylonitrile	Ammonium persulfate	FTIR, DSC	[56]
18	Starch-g-poly(acrylic acid)	Acrylic acid	Ceric ammonium nitrate, potassium persulfate	FTIR, TGA, SEM	[17]
19	AAm-co-MD	Acrylamide	Ammonium persulphate	FTIR, 1H-NMR	[58]
20	Poly AA–co-AM clay	Acrylic acid, acrylamide	Ammonium persulfate	FTIR, XRD, SEM, TGA	[37]

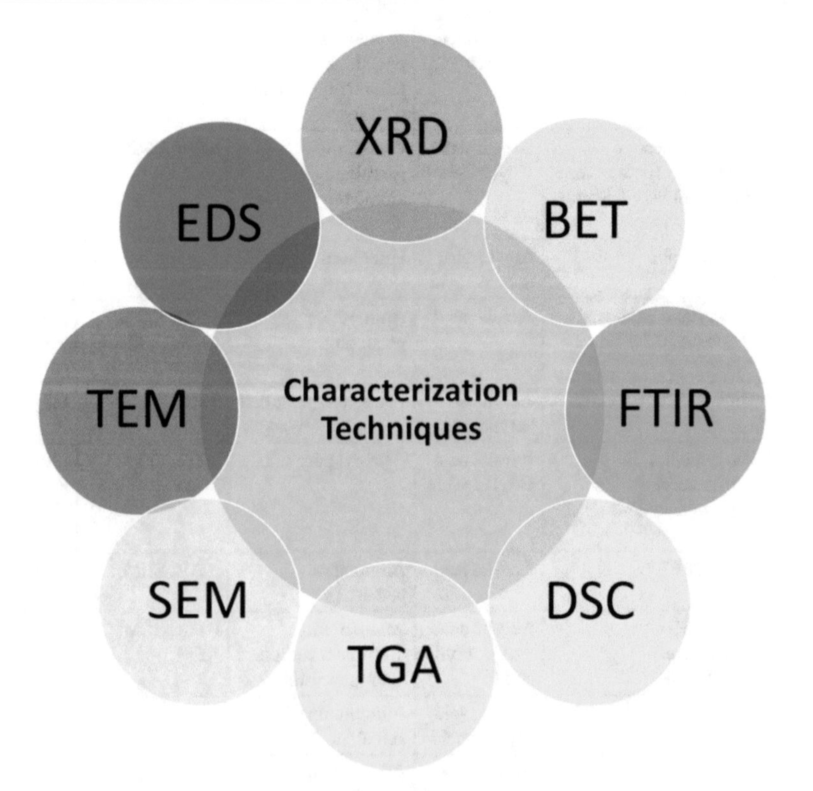

Fig. 10 Characterization techniques for bio-based superabsorbents

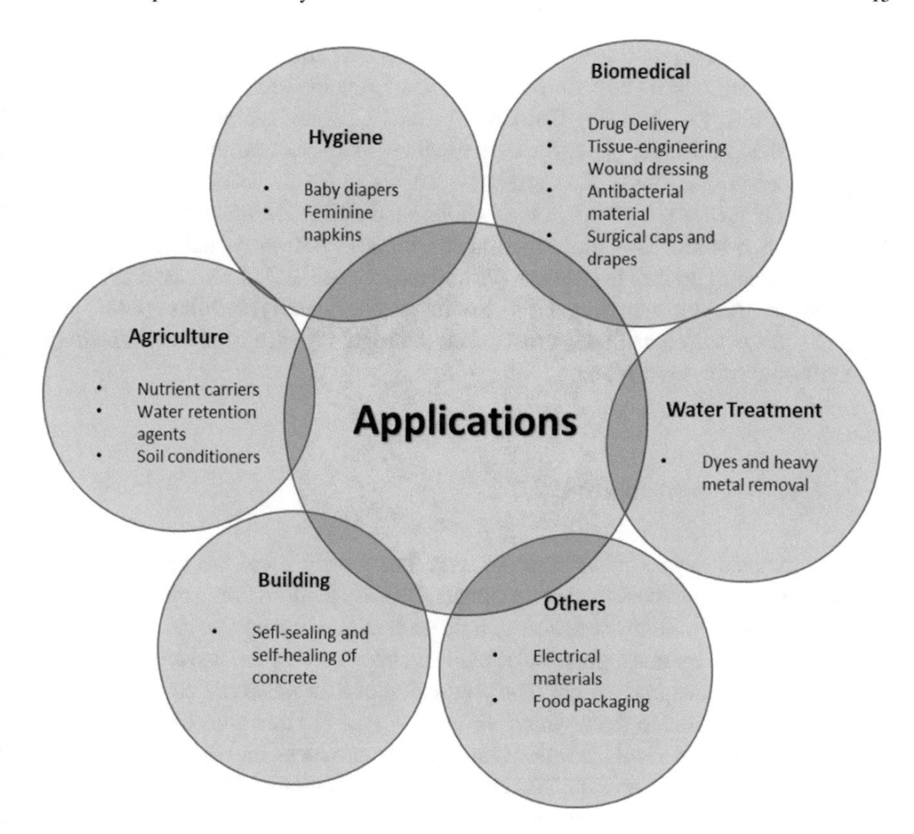

Fig. 11 Applications of superabsorbents in different fields

simple to administer medications through the peroral, vaginal, rectal, ocular, and transdermal routes when SAs used in gel form. SAs based on cellulose, alginate, and hyaluronic acid are being utilized in wound healing and other biomedical applications. Bio-based SAs are biodegradable and biocompatible in nature and these are being examined for their antibacterial characteristics. SAs based on chitosan, exhibit robust biocidal act against specific bacteria species because of the positive charge of quaternary ammoniums, which attracts the negative charge from microbial membranes and damages the cell membrane [7, 26, 62, 85].

5.2 *Water Treatment Applications*

One of the most significant environmental problem faced by civilization today is water contamination. WHO estimates that this issue, particularly in underdeveloped nations, causes about 14,000 deaths every day. The industrial revolution brings a serious crisis of water pollution which impact humans, animals, environment, and aquatic life. Wastewater generated from a variety of industries contains synthetic dyes and harmful heavy metal ions in sizeable-capacities. Heavy metals like Zn, Cu, Ni, Hg, Co, Cd, Pb, and Cr, and synthetic dyes such as crystal violet, malachite green, and methylene blue are mainly responsible for water pollution. Bio-based SAs interact with water toxins and bind these toxins through their functional groups, for example, –COOH, –OH, and –NH2 and purify the water. Bio-based SAs like crosslinked chitosan-graft-poly(maleic acid) [28], poly(acrylic acid) grafted to chitosan [75], Pineapple peel cellulose-based hydrogels [20], Chitosan/cellulose hydrogel [99], Sodium carboxymethyl cellulose (CMC-Na)/ sodium styrene sulfonate (SSS) crosslinked hydrogel [98] are used to eliminate the contaminants from wastewater.

5.3 Hygiene Applications

Superabsorbents have a wide range of uses, but mostly these are utilized for the manufacturing of personal hygiene products. SAs are the functional elements found inside sanitary towels and diapers. Any fluid such as blood and urine can be absorbed and held in place by these powerful chemical substances in the capacity of up to a hundred times its original weight. The production of SAs was more than one million tons in 1990 worldwide from which 81% were used for newborn diapers, 8% for adult incontinence products, 5% used for feminine products, and the rest for various medical uses. Initially, synthetic SAs were used to produce such hygiene products but an enlightenment about environmental issues due to synthetic materials leads to the inspiration for synthesizing bio-based superabsorbents. Consequently, contemporary industry is now concerned with the search for bio-disposable superabsorbents [16, 75]. An example of bio-based superabsorbents used for the production of hygiene products is crosslinked sodium carboxymethyl cellulose and hydroxyethyl cellulose using divinyl sulfone [84].

5.4 Agricultural Applications

The most damaging factor for crops is drought, which inhibits development, reduces biomass production, and, most significantly, slows down photosynthesis. The biogeochemical P cycle is also disrupted by a water shortage, which affects the nutrients available to plants or crops. These may disrupt the productivity of essential plants or crops causing financial damage. With the help of polymer chemistry, these agricultural problems can be eliminated. SAs became the center of attention in the scientific community in recent decades due to their larger water absorption capability, nontoxic, biocompatibility, photostability, high durability, stability on swelling, long storage of effluent, biodegradability, stability on wide temperature range, colorless, odorless, improve soil quality in chemical, physical, and biological manner, rewetting capability for a longer time, and cost-effectiveness make it highly applicable in agriculture sector. According to reports, hydrogels can help soils retain more water or fertilizer and release them on need in a slow and constant manner [3, 23, 40, 66, 74]. Examples of SAs like whey/cellulose-based hydrogel prepared from acid crosslinking of hydroxyethyl cellulose and carboxymethyl cellulose [25], Lignin-based SAs prepared using poly(ethylene glycol) diglycidyl ether and lignin alkali polymers [59], and cellulose acetate- based SAs crosslinked prepared using EDTA; [90] are utilized for irrigation and fertilizer release for effective growth of crops.

5.5 *Construction Applications*

In past years, researchers investigated various SAs to be used in concrete and building materials. These techniques ordinarily deal with self-sealing and self-healing in fractures of concrete. The self-healing properties of these materials help in fixing the cracks and fractures in the building materials without the external support and help of humans. These materials thus decrease the building expenditure. Concrete can suffer from autogenous shrinkage because of the water. With the help of SAs, these problems of premature cracks due to autogenous shrinkage can be lessened. In order to harden the concrete and minimize the fracture and cracks, different bio-based SAs are being prepared by the research community. Materials such as xanthan gum, agarose, alginate, chitosan and carrageenan are used to prepare SAs which serve the same purpose. For researchers, prepared bio-based SAs prepared using poly(aspartic acid) [22], dextrin-polyacrylamide and boric acid [76], acrylamide/acrylic acid [88], and κ-carrageenan/poly(acrylic acid) [2] have properties of self-healing.

5.6 *Other Applications*

SAs can be utilized in different fields such as electrical purposes and food packaging industries. In the electrical field, SAs can be used as coating and tapes for preventing the power cables from water and moisture; starch-grafted sodium polyacrylates-based SAs are already being used in the electrical field [75]. Additionally, SAs powders are employed in the preservation of hydrophilic solutions in food packaging. Cross-linked carboxymethyl cellulose is used for preparing SAs powder and supportive materials to hold fluids in the package which was developed [114].

Fig. 12 Top 100 words in the abstract of research publications on bio-based SAs

6 Bibliometric Analysis

6.1 Abstract Analysis of Different Research Articles on Bio-based Superabsorbents

Figure 12 indicated the top 100 words in the word cloud prepared from abstracts of research publications related to bio-based superabsorbents. Words "hydrogels", "bio-based", "materials", "properties", "water", "applications", "chemical", "polymers", "drug", and "cellulose" stands in the top 10 words of the title analysis of this study. "Hydrogel" topped the list of 100 words and was used 189 times which is quite obvious. Careful observation of the words analysis study indicated the importance of words such as "hydrogels", "bio-based", "drug", "cellulose", "pH", and "polymer" in bio-based SAs. "Compared", "conditions", "including", and "increase" are the least used words in the abstract of various research articles and all are used only 15 times.

6.2 Year Wise Publications

A total of 90 research papers were published in the period of 2011–2021 on the topic of bio-based superabsorbents, data collected from the Scopus database. In the year 2021, maximum publications were published, a total of 25 papers were published which was followed by the year 2019 with 17 research publications. 01 is the least number of research papers that were published in the year 2013. Current analysis revealed that "3", "1", "5", "7", "8", "5", "6", "17", "13", and "25" research papers published in 2011, 2012, 2013, 2014, 2015, 2016, 2017, 2018, 2019, 2020, and 2021, respectively (Fig. 13).

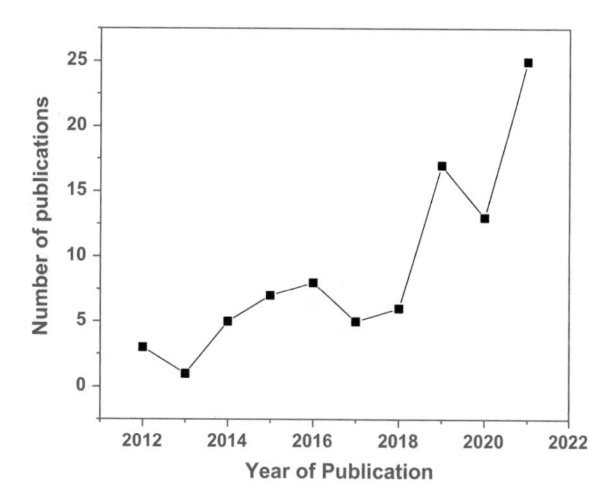

Fig. 13 Year wise trend of publications on bio-based SAs

6.3 *Keyword Analysis*

Figure 14 shows the analysis of keywords presented by researchers in their research publications on bio-based superabsorbents by counting method. Top ten words are "hydrogel", "bio-based", "drug", "chitosan", "delivery", "cross-linking", "acid", "applications", "polymer", and "cellulose". In the word cloud (Fig. 14), the larger the size of a word is, the greater its occurrence in the research articles. Analysis of these keywords revealed that bio-based materials such as chitosan are leading material used for the synthesis of SAs.

Fig. 14 Analysis of keywords in research publications on the synthesis of bio-based SAs

Fig. 15 Analysis of titles in research publications on the synthesis of bio-based SAs

6.4 Title Analysis

Figure 15 indicated the top 100 words in the word cloud prepared from titles of research publications related to dye adsorption by anion exchangers. Words "hydrogels", "bio-based", "synthesis", "chitosan", "applications", "properties", "based", "polymers", "materials", and "alginate" stands in the top 10 words of the title analysis of this study. "Hydrogels" topped the list of 100 words and was used 33 times which is quite obvious. A close look at the words analysis study indicated the importance of words such as "hydrogels", "delivery", "materials", "cellulose", "drug", and "bio-based" in SAs synthesis. "Prepared", "pH-responsive", "platform", "polysaccharide", "reactive", "nanoparticles", "nano-crystals", "modifiers", "independent", "flow", "florescence", "formulation", "gels", "Nano-chitin", "microfibers", "methacrylate", "mechanical", "itaconic", "improved", "imaging", "functional", "advanced", "capacity", "autogenous", "capacity", "chemicals", "medical", "chain", "biomedical", "biomaterials", "antibacterial", "amphiphilic", "natural", "nanocomposite", "pectin-based", etc., are least used words in titles of various research articles and are used only 2 times each.

References

1. Abdollahi, Z., Frounchi, M., Dadbin, S.: Synthesis, characterization and comparison of PAM, cationic PDMC and P(AM-co-DMC) based on solution polymerization. J. Ind. Eng. Chem. **17**(3), 580–586 (2011). https://doi.org/10.1016/j.jiec.2010.10.030
2. Aday, A.N., Osio-Norgaard, J., Foster, K.E.O., Srubar, W.V.: Carrageenan-based superabsorbent biopolymers mitigate autogenous shrinkage in ordinary Portland cement. Mater. Struct. **51**(2), 37 (2018). https://doi.org/10.1617/s11527-018-1164-5
3. Adjuik, T.A., Nokes, S.E., Montross, M.D., Wendroth, O.: The impacts of bio-based and synthetic hydrogels on soil hydraulic properties: a review. In: Polymers, vol. 14, issue 21 (2022). https://doi.org/10.3390/polym14214721

4. Alam, M.N., Christopher, L.P.: Natural cellulose-chitosan cross-linked superabsorbent hydrogels with superior swelling properties. ACS Sustain. Chem. Eng. **6**(7), 8736–8742 (2018). https://doi.org/10.1021/acssuschemeng.8b01062
5. An, T.B., Huynh, D., Linh, T., Van, N.T.T., Tien, H.V., Anh, P., Tri, N.: Synthesis and characterization of salt-resistant super absorbent derived from maize bran **11**, 120–124 (2020)
6. Ashfaq, A., Clochard, M.-C., Coqueret, X., Dispenza, C., Driscoll, M.S., Ulański, P., Al-Sheikhly, M.: Polymerization reactions and modifications of polymers by ionizing radiation. Polymers **12**(12) (2020). https://doi.org/10.3390/polym12122877
7. Aziz, M.A., Cabral, J.D., Brooks, H.J.L., Moratti, S.C., Hanton, L.R.: Antimicrobial properties of a chitosan dextran-based hydrogel for surgical use. Antimicrob. Agents Chemother. **56**(1), 280–287 (2012). https://doi.org/10.1128/AAC.05463-11
8. Bachra, Y., Grouli, A., Damiri, F., Bennamara, A., Berrada, M.: A new approach for assessing the absorption of disposable baby diapers and superabsorbent polymers: a comparative study. Results Mater. **8**, 100156 (2020). https://doi.org/10.1016/j.rinma.2020.100156
9. Bhat, G., Kandagor, V.: 1—Synthetic polymer fibers and their processing requirements (D. B. T.-A. in F. Y. S. of T. and P. Zhang (ed.), pp. 3–30). Woodhead Publishing (2014). https://doi.org/10.1533/9780857099174.1.3
10. Braihi, A: Applications of the Super Absorbent Polymers. ResearchGate, pp. 1–4 (2016)
11. Brydson, J.A.: 15—Acrylic Plastics (J. A. B. T.-P. M. (Seventh E. Brydson (ed.), pp. 398–424). Butterworth-Heinemann (1999). https://doi.org/10.1016/B978-075064132-6/50056-5
12. Capezza, A.J., Newson, W.R., Olsson, R.T., Hedenqvist, M.S., Johansson, E.: Advances in the use of protein-based materials: toward sustainable naturally sourced absorbent materials. ACS Sustain. Chem. Eng. **7**(5), 4532–4547 (2019). https://doi.org/10.1021/acssuschemeng.8b05400
13. Chang, L., Xu, L., Liu, Y., Qiu, D.: Superabsorbent polymers used for agricultural water retention. Polym. Test. **94**, 107021 (2021). https://doi.org/10.1016/j.polymertesting.2020.107021
14. Chapiro, A.: Radiation induced polymerization. Radiat. Phys. Chem. (1977) **14**(1), 101–116 (1979). https://doi.org/10.1016/0146-5724(79)90015-3
15. Chen, M., Shen, Y., Xu, L., Xiang, G., Ni, Z.: Synthesis of a super-absorbent nanocomposite hydrogel based on vinyl hybrid silica nanospheres and its properties. RSC Adv. **10**(67), 41022–41031 (2020). https://doi.org/10.1039/D0RA07074B
16. Cheng, B., Pei, B., Wang, Z., Hu, Q.: Advances in chitosan-based superabsorbent hydrogels. RSC Adv. **7**(67), 42036–42046 (2017). https://doi.org/10.1039/C7RA07104C
17. Czarnecka, E., Nowaczyk, J.: Semi-natural superabsorbents based on starch-g-poly(acrylic acid): modification, synthesis and application. In: Polymers, vol. 12, issue 8 (2020). https://doi.org/10.3390/polym12081794
18. Czarnecka, E., Nowaczyk, J.: Synthesis and characterization superabsorbent polymers made of starch, acrylic acid, acrylamide, poly(Vinyl alcohol), 2-hydroxyethyl methacrylate, 2-acrylamido-2-methylpropane sulfonic acid. Int. J. Mol. Sci. **22**(9) (2021). https://doi.org/10.3390/ijms22094325
19. Dabbaghi, A., Rahmani, S.: Synthesis and characterization of biodegradable multicomponent amphiphilic conetworks with tunable swelling through combination of ring-opening polymerization and "click" chemistry method as a controlled release formulation for 2,4-dichlorophenoxyacetic a. Polym. Adv. Technol. **30**(2), 368–380 (2019). https://doi.org/10.1002/pat.4474
20. Dai, H., Huang, H.: Synthesis, characterization and properties of pineapple peel cellulose-g-acrylic acid hydrogel loaded with kaolin and sepia ink. Cellulose **24**(1), 69–84 (2017). https://doi.org/10.1007/s10570-016-1101-0
21. Das, A., Kumar, A., Patil, N.B., Viswanathan, C., Ghosh, D.: Preparation and characterization of silver nanoparticle loaded amorphous hydrogel of carboxymethylcellulose for infected wounds. Carbohydr. Polym. **130**, 254–261 (2015). https://doi.org/10.1016/j.carbpol.2015.03.082

22. De Grave, L., Tenório Filho, J.R., Snoeck, D., Vynnytska, S., De Belie, N., Bernaerts, K.V., Van Vlierberghe, S.: Poly(aspartic acid) superabsorbent polymers as biobased and biodegradable additives for self-sealing of cementitious mortar. J. Sustain. Cement-Based Mater. 1–16 (2022). https://doi.org/10.1080/21650373.2022.2137861
23. Ding, C., Zhang, S., Fu, X., Liu, T., Shao, L., Fei, M., Hao, C., Liu, Y., Zhong, W.-H.: Robust supramolecular composite hydrogels for sustainable and "visible" agriculture irrigation. J. Mater. Chem. A **9**(43), 24613–24621 (2021). https://doi.org/10.1039/D1TA05442B
24. Ding, X., Li, L., Liu, P., Zhang, J., Zhou, N., Lu, S., Wei, S., Shen, J.: The preparation and properties of dextrin-graft-acrylic acid/montmorillonite superabsorbent nanocomposite. Polym. Compos. **30**(7), 976–981 (2009). https://doi.org/10.1002/pc.20643
25. Durpekova, S., Filatova, K., Cisar, J., Ronzova, A., Kutalkova, E., Sedlarik, V.: A novel hydrogel based on renewable materials for agricultural application. Int. J. Polym. Sci. **2020**, 8363418 (2020). https://doi.org/10.1155/2020/8363418
26. Fu, L.-H., Qi, C., Ma, M.-G., Wan, P.: Multifunctional cellulose-based hydrogels for biomedical applications. J. Mater. Chem. B **7**(10), 1541–1562 (2019). https://doi.org/10.1039/C8TB02331J
27. Garner, C.M., Nething, M., Nguyen, P.: The synthesis of a superabsorbent polymer. J. Chem. Educ. **74**(1), 95 (1997). https://doi.org/10.1021/ed074p95
28. Ge, H., Hua, T.: Synthesis and characterization of poly(maleic acid)-grafted crosslinked chitosan nanomaterial with high uptake and selectivity for Hg(II) sorption. Carbohydr. Polym. **153**, 246–252 (2016). https://doi.org/10.1016/j.carbpol.2016.07.110
29. Gooch, J.W.: Superabsorbent Polymers BT—Encyclopedic Dictionary of Polymers (J. W. Gooch (ed.), pp. 714–715). Springer New York (2011). https://doi.org/10.1007/978-1-4419-6247-8_11417
30. Goponenko, A.V., Dzenis, Y.A.: Role of mechanical factors in applications of stimuli-responsive polymer gels—status and prospects. Polymer **101**, 415–449 (2016). https://doi.org/10.1016/j.polymer.2016.08.068
31. Green, M.F., Bisby, L.A., Beaudoin, Y., Labossière, P.: Effect of freeze-thaw cycles on the bond durability between fibre reinforced polymer plate reinforcement and concrete. Can. J. Civ. Eng. **27**(5), 949–959 (2000). https://doi.org/10.1139/l00-031
32. Guan, H., Li, J., Zhang, B., Yu, X.: Synthesis, properties, and humidity resistance enhancement of biodegradable cellulose-containing superabsorbent polymer. J. Polym. **2017**, 3134681 (2017). https://doi.org/10.1155/2017/3134681
33. Gupta, S., Goswami, S., Sinha, A.: A combined effect of freeze–thaw cycles and polymer concentration on the structure and mechanical properties of transparent PVA gels. Biomed. Mater. (bristol, England) **7**(1), 15006 (2012). https://doi.org/10.1088/1748-6041/7/1/015006
34. He, M., Zhao, Y., Duan, J., Wang, Z., Chen, Y., Zhang, L.: Fast contact of solid-liquid interface created high strength multi-layered cellulose hydrogels with controllable size. ACS Appl. Mater. Interfaces **6**(3), 1872–1878 (2014). https://doi.org/10.1021/am404855q
35. Hedir, G.G., Bell, C.A., O'Reilly, R.K., Dove, A.P.: Functional degradable polymers by radical ring-opening copolymerization of MDO and vinyl bromobutanoate: synthesis degradability and post-polymerization modification. Biomacromol **16**(7), 2049–2058 (2015). https://doi.org/10.1021/acs.biomac.5b00476
36. Hoecker, F.E., Watkins, I.: Radiation polymerization of liquids. Radiology **68**(2), 257 (1957). https://doi.org/10.1148/68.2.257a
37. Iqbal, M., Ahmed, I., Saeed Khan, M., Salman, S.M., Noor, S.: Preparation and characterization of co-polymer (acrylic acid and acrylamide) as super absorbent composites grafted with Thar clay. Sci. Int. (lahore) **32**(4), 2020 (2020)
38. Itsubo, N., Wada, M., Imai, S., Myoga, A., Makino, N.: Life Cycle Assessment of the Closed-Loop Recycling, pp. 1–15 (2020)
39. Jafari, M., Najafi, G.R., Sharif, M.A., Elyasi, Z.: Superabsorbent polymer composites derived from polyacrylic acid: design and synthesis, characterization, and swelling capacities. Polym. Polym. Compos. **29**(6), 733–739 (2020). https://doi.org/10.1177/0967391120933482

40. Kaur, R., Sharma, R., Chahal, G.K.: Synthesis of lignin-based hydrogels and their applications in agriculture: a review. Chem. Pap. **75**(9), 4465–4478 (2021). https://doi.org/10.1007/s11696-021-01712-w
41. Khanlari, S., Dubé, M.A.: Effect of pH on poly(acrylic acid) solution polymerization. J. Macromol. Sci. Part A **52**(8), 587–592 (2015). https://doi.org/10.1080/10601325.2015.1050628
42. Klinpituksa, P., Kosaiyakanon, P.: Superabsorbent polymer based on sodium carboxymethyl cellulose grafted polyacrylic acid by inverse suspension polymerization. Int. J. Polym. Sci. **2017**, 3476921 (2017). https://doi.org/10.1155/2017/3476921
43. Kong, W., Li, Q., Liu, J., Li, X., Zhao, L., Su, Y., Yue, Q., Gao, B.: Adsorption behavior and mechanism of heavy metal ions by chicken feather protein-based semi-interpenetrating polymer networks super absorbent resin. RSC Adv. **6**(86), 83234–83243 (2016). https://doi.org/10.1039/C6RA18180E
44. Kontopoulou, I., Congdon, T.R., Bassett, S., Mair, B., Gibson, M.I.: Synthesis of poly(vinyl alcohol) by blue light bismuth oxide photocatalysed RAFT. Evaluation of the impact of freeze/thaw cycling on ice recrystallisation inhibition. Polym. Chem. **13**(32), 4692–4700 (2022). https://doi.org/10.1039/d2py00852a
45. Korpe, S., Erdoğan, B., Bayram, G., Ozgen, S., Uludag, Y., Bicak, N.: Crosslinked DADMAC polymers as cationic super absorbents. Reactive Funct. Polym. **69**(9), 660–665 (2009). https://doi.org/10.1016/j.reactfunctpolym.2009.04.010
46. Kurt, B., Gulyuz, U., Demir, D.D., Okay, O.: High-strength semi-crystalline hydrogels with self-healing and shape memory functions. Eur. Polym. J. **81**, 12–23 (2016). https://doi.org/10.1016/j.eurpolymj.2016.05.019
47. Ladelta, V., Bilalis, P., Gnanou, Y., Hadjichristidis, N.: Ring-opening polymerization of ω-pentadecalactone catalyzed by phosphazene superbases. Polym. Chem. **8**(3), 511–515 (2017). https://doi.org/10.1039/C6PY01983H
48. Laitinen, O., Suopajärvi, T., Österberg, M., Liimatainen, H.: Hydrophobic, superabsorbing aerogels from choline chloride-based deep eutectic solvent pretreated and silylated cellulose nanofibrils for selective oil removal. ACS Appl. Mater. Interfaces **9**(29), 25029–25037 (2017). https://doi.org/10.1021/acsami.7b06304
49. Liu, H., Yu, M., Ma, H., Wang, Z., Li, L., Li, J.: Pre-irradiation induced emulsion co-graft polymerization of acrylonitrile and acrylic acid onto a polyethylene nonwoven fabric. Radiation Phys. Chem. **94**, 129–132 (2014). https://doi.org/10.1016/j.radphyschem.2013.06.023
50. Liu, Z., Miao, Y., Wang, Z., Yin, G.: Synthesis and characterization of a novel super-absorbent based on chemically modified pulverized wheat straw and acrylic acid. Carbohydr. Polym. **77**(1), 131–135 (2009). https://doi.org/10.1016/j.carbpol.2008.12.019
51. Llanes, L., Dubessay, P., Pierre, G., Delattre, C., Michaud, P.: Biosourced Polysaccharide-Based Superabsorbents, pp. 51–79 (2020)
52. Lodhi, B.A., Hussain, M.A., Sher, M., Haseeb, M.T., Ashraf, M.U., Hussain, S.Z., Hussain, I., Bukhari, S.N.A.: Polysaccharide-based superporous, superabsorbent, and stimuli responsive hydrogel from sweet basil: a novel material for sustained drug release. Adv. Polym. Technol. **2019**, 9583516 (2019). https://doi.org/10.1155/2019/9583516
53. Lv, N., Wang, X., Peng, S., Luo, L., Zhou, R.: Superhydrophobic/superoleophilic cotton-oil absorbent: preparation and its application in oil/water separation. RSC Adv. **8**(53), 30257–30264 (2018). https://doi.org/10.1039/C8RA05420G
54. Ma, X., Wen, G.: Development history and synthesis of super-absorbent polymers: a review. J. Polym. Res. **27**(6) (2020). https://doi.org/10.1007/s10965-020-02097-2
55. Madhav, H., Singh, N., Jaiswar, G.: Chapter 4—Thermoset, bioactive, metal–polymer composites for medical applications (V. Grumezescu & A. M. B. T.-M. for B. E. Grumezescu (eds.), pp. 105–143) (2019). Elsevier. https://doi.org/10.1016/B978-0-12-816874-5.00004-9
56. Mahatmanti, F.W., Harjono, Assa'Idah, I.N.: Synthesis and characterization of chitosan based super absorbent polymer modified with acrylic acid and acrylonitrile for Pb (II) metal ions removal from water. AIP Conf. Proc. **2237**(Ii) (2020). https://doi.org/10.1063/5.0005748

57. Marques, D.A.S., Jarmelo, S., Baptista, C.M.S.G., Gil, M.H.: Poly(lactic acid) synthesis in solution polymerization. Macromol. Symp. **296**(1), 63–71 (2010). https://doi.org/10.1002/masy.201051010
58. Mazi, H., Surmelihindi, B.: Temperature and ph-sensitive super absorbent polymers based on modified maleic anhydride. J. Chem. Sci. **133**(1) (2021). https://doi.org/10.1007/s12039-020-01873-3
59. Mazloom, N., Khorassani, R., Zohuri, G.H., Emami, H., Whalen, J.: Development and characterization of lignin-based hydrogel for use in agricultural soils: preliminary evidence. CLEAN—Soil, Air, Water **47**(11), 1900101 (2019). https://doi.org/10.1002/clen.201900101
60. McLaughlin, J.R., Abbott, N.L., Guymon, C.A.: Responsive superabsorbent hydrogels via photopolymerization in lyotropic liquid crystal templates. Polymer **142**, 119–126 (2018). https://doi.org/10.1016/j.polymer.2018.03.016
61. Mechtcherine, V., Wyrzykowski, M., Schröfl, C., Snoeck, D., Lura, P., De Belie, N., Mignon, A., Van Vlierberghe, S., Klemm, A.J., Almeida, F.C.R., Tenório Filho, J.R., Boshoff, W.P., Reinhardt, H.-W., Igarashi, S.-I.: Application of super absorbent polymers (SAP) in concrete construction—update of RILEM state-of-the-art report. Mater. Struct. **54**(2), 80 (2021). https://doi.org/10.1617/s11527-021-01668-z
62. Mihajlovic, M., Fermin, L., Ito, K., Van Nostrum, C.F., Vermonden, T.: Hyaluronic acid-based supramolecular hydrogels for biomedical applications. Multifunctional Mater. **4**(3) (2021). https://doi.org/10.1088/2399-7532/ac1c8a
63. Miyazaki, T., Yamaoka, K., Gong, J.P., Osada, Y.: Hydrogels with crystalline or liquid crystalline structure. Macromol. Rapid Commun. **23**(8), 447–455 (2002). https://doi.org/10.1002/1521-3927(20020501)23:8<447::AID-MARC447>3.0.CO;2-O
64. Mudiyanselage, T.K., Neckers, D.C.: Highly absorbing superabsorbent polymer. J. Polym. Sci. Part A: Polym. Chem. **46**(4), 1357–1364 (2008). https://doi.org/10.1002/pola.22476
65. Mujumdar, A.N., Young, S.G., Merker, R.L., Magill, J.H.: A study of solution polymerization of polyphosphazenes. Macromolecules **23**(1), 14–21 (1990). https://doi.org/10.1021/ma00203a004
66. Naidu, D.S., Hlangothi, S.P., John, M.J.: Bio-based products from xylan: a review. Carbohydr. Polym. **179**, 28–41 (2018). https://doi.org/10.1016/j.carbpol.2017.09.064
67. Nakano, T., Nakaoki, T.: Coagulation size of freezable water in poly(vinyl alcohol) hydrogels formed by different freeze/thaw cycle periods. Polym. J. **43**(11), 875–880 (2011). https://doi.org/10.1038/pj.2011.92
68. Narita, T., Mayumi, K., Ducouret, G., Hébraud, P.: Viscoelastic properties of poly(vinyl alcohol) hydrogels having permanent and transient cross-links studied by microrheology, classical rheometry, and dynamic light scattering. Macromolecules **46**(10), 4174–4183 (2013). https://doi.org/10.1021/ma400600f
69. Okay, O.: Semicrystalline physical hydrogels with shape-memory and self-healing properties. J. Mater. Chem. B **7**(10), 1581–1596 (2019). https://doi.org/10.1039/C8TB02767F
70. Oladosu, Y., Rafii, M.Y., Arolu, F., Chukwu, S.C., Salisu, M.A., Fagbohun, I.K., Muftaudeen, T.K., Swaray, S., Haliru, B.S.: Superabsorbent polymer hydrogels for sustainable agriculture: a review. Horticulturae **8**(7), 1–17 (2022). https://doi.org/10.3390/horticulturae8070605
71. Ostrand, M.S., DeSutter, T.M., Daigh, A.L.M., Limb, R.F., Steele, D.D.: Superabsorbent polymer characteristics, properties, and applications. Agrosyst. Geosci. Environ. **3**(1), 1–14 (2020). https://doi.org/10.1002/agg2.20074
72. Padsalgikar, A.D.: 1—Introduction to plastics. In A. D. B. T.-P. in M. D. for C. A. Padsalgikar (ed.) Plastics Design Library, pp. 1–29. William Andrew Publishing (2017). https://doi.org/10.1016/B978-0-323-35885-9.00001-1
73. Parvathy, P.C., Jyothi, A.N.: Rheological and thermal properties of saponified cassava starch-g-poly(acrylamide) superabsorbent polymers varying in grafting parameters and absorbency. J. Appl. Polym. Sci. **131**(11) (2014). https://doi.org/10.1002/app.40368
74. Patra, S.K., Poddar, R., Brestic, M., Acharjee, P.U., Bhattacharya, P., Sengupta, S., Pal, P., Bam, N., Biswas, B., Barek, V., Ondrisik, P., Skalicky, M., Hossain, A.: Prospects of hydrogels in agriculture for enhancing crop and water productivity under water deficit condition. Int. J. Polym. Sci. **2022**, 4914836 (2022). https://doi.org/10.1155/2022/4914836

75. Pérez-Álvarez, L., Ruiz-Rubio, L., Lizundia, E., Vilas-Vilela, J.L.: Polysaccharide-based superabsorbents: synthesis, properties, and applications BT—cellulose-based superabsorbent hydrogels (M. I. H. Mondal (ed.), pp. 1393–1431). Springer International Publishing (2019). https://doi.org/10.1007/978-3-319-77830-3_46
76. Priya, Sharma, A.K., Kaith, B.S., Simran, Bhagyashree, Arora, S.: Synthesis of dextrin-polyacrylamide and boric acid based tough and transparent, self-healing, superabsorbent film. Int. J. Biol. Macromol. **182**, 712–721 (2021). https://doi.org/10.1016/j.ijbiomac.2021.04.028
77. Range, K.M.D., Moser, Y.A.: NIH public access. Bone, **23**(1), 1–7 (2012). https://doi.org/10.1039/c3cs35512h.Stimuli-responsive
78. Rather, R.A., Bhat, M.A., Shalla, A.H.: An insight into synthetic and physiological aspects of superabsorbent hydrogels based on carbohydrate type polymers for various applications: a review. Carbohydr. Polym. Technol. Appl. **3**, 100202 (2022). https://doi.org/10.1016/j.carpta.2022.100202
79. Rehman, T.U., Shah, L.A., Khan, M., Irfan, M., Khattak, N.S.: Zwitterionic superabsorbent polymer hydrogels for efficient and selective removal of organic dyes. RSC Adv. **9**(32), 18565–18577 (2019). https://doi.org/10.1039/C9RA02488C
80. Riswati, S.S., Setiati, R., Kasmungin, S., Prakoso, S., Fathaddin, M.T.: Sugarcane bagasse for environmentally friendly super-absorbent polymer: Synthesis methods and potential applications in oil industry. IOP Conf. Ser. Earth Environ. Sci. **819**(1) (2021). https://doi.org/10.1088/1755-1315/819/1/012017
81. Rooney, J.M.: 43—Carbocationic polymerization: N-vinylcarbazole (G. Allen & J. C. B. T.-C. P. S. and S. Bevington (eds.) pp. 697–704). Pergamon (1989). https://doi.org/10.1016/B978-0-08-096701-1.00105-1
82. Rosales, A.M., Anseth, K.S.: The design of reversible hydrogels to capture extracellular matrix dynamics. Nat. Rev. Mater. **1** (2016). https://doi.org/10.1038/natrevmats.2015.12
83. Rosales, A.M., Anseth, K.S.: The design of reversible hydrogels to capture extracellular matrix dynamics. Nat. Rev. Mater. **1**(2), 15012 (2016). https://doi.org/10.1038/natrevmats.2015.12
84. Sannino, A., Demitri, C., Madaghiele, M.: Biodegradable Cellulose-based Hydrogels: Design and Applications. In: Materials, vol. 2, issue 2, pp. 353–373 (2009). https://doi.org/10.3390/ma2020353
85. Saptaji, K., Iza, N.R., Widianingrum, S., Mulia, V.K., Setiawan, I.: Poly(2-hydroxyethyl methacrylate) hydrogels for contact lens applications–a review. Makara J. Sci. **25**(3), 145–154 (2021). https://doi.org/10.7454/mss.v25i3.1237
86. Sarkar Phyllis, A.K., Tortora, G., Johnson, I.: Super-Absorbent Polymer. The Fairchild Books Dictionary of Textiles (2022). https://doi.org/10.5040/9781501365072.15873
87. Sastri, V.R.: 5—Commodity thermoplastics: polyvinyl chloride, polyolefins, cycloolefins and polystyrene. In: V. R. B. T.-P. in M. D. (Third E. Sastri (ed.) Plastics Design Library, pp. 113–166. William Andrew Publishing (2022). https://doi.org/10.1016/B978-0-323-85126-8.00002-3
88. Schreiberová, H., Fládr, J., Trtík, T., Chylík, R., Bílý, P.: An investigation of the compatibility of different approaches to self-healing concrete: the superabsorbent polymers and microbially induced calcite precipitation. IOP Conf. Ser. Mater. Sci. Eng. **596**(1), 12039 (2019). https://doi.org/10.1088/1757-899X/596/1/012039
89. Schröfl, C.: Superabsorbent polymers. In: Van Vlierberghe, S., Mignon, A. (eds.) Chemical Design, Processing and Applications, pp. 1–34. De Gruyter (n.d.). https://doi.org/10.1515/9781501519116-002
90. Senna, A.M., Botaro, V.R.: Biodegradable hydrogel derived from cellulose acetate and EDTA as a reduction substrate of leaching NPK compound fertilizer and water retention in soil. J. Controlled Release Official J. Controlled Release Society **260**, 194–201 (2017). https://doi.org/10.1016/j.jconrel.2017.06.009
91. Sharma, S., Dua, A., Malik, A.: Biocompatible stimuli responsive superabsorbent polymer for controlled release of GHK-Cu peptide for wound dressing application. J. Polym. Res. **24**(7), 104 (2017). https://doi.org/10.1007/s10965-017-1254-z

92. Shin, B.M., Kim, J.-H., Chung, D.J.: Synthesis of pH-responsive and adhesive super-absorbent hydrogel through bulk polymerization. Macromol. Res. **21**(5), 582–587 (2013). https://doi.org/10.1007/s13233-013-1051-4
93. Soleimani, F., Sadeghi, M., Shasevari, H., Soleimani, A., Sadeghi, H.: Synthesis of super absorbent hydrogels with using releasing drug. Asian J. Chem. **25**(2), 1025–1028 (2013). https://doi.org/10.14233/ajchem.2013.13394
94. Soto, D., Urdaneta, J., Pernía, K., León, O., Muñoz-Bonilla, A., Fernández-García, M.: Heavy metal (Cd2+, Ni2+, Pb2+ and Ni2+) adsorption in aqueous solutions by oxidized starches. Polym. Adv. Technol. **26**(2), 147–152 (2015). https://doi.org/10.1002/pat.3439
95. Stauffer, S.R., Peppast, N.A.: Poly(vinyl alcohol) hydrogels prepared by freezing-thawing cyclic processing. Polymer **33**(18), 3932–3936 (1992). https://doi.org/10.1016/0032-3861(92)90385-A
96. Tang, H., Luan, Y., Yang, L., Sun, H.: A perspective on reversibility in controlled polymerization systems: recent progress and new opportunities. In: Molecules, vol. 23, issue 11 (2018). https://doi.org/10.3390/molecules23112870
97. Tinajero-Díaz, E., Kimmins, S.D., García-Carvajal, Z.-Y., Martínez de Ilarduya, A.: Polypeptide-based materials prepared by ring-opening polymerisation of anionic-based α-amino acid N-carboxyanhydrides: a platform for delivery of bioactive-compounds. Reactive Funct. Polym. **168**, 105040 (2021). https://doi.org/10.1016/j.reactfunctpolym.2021.105040
98. Tran, T.H., Okabe, H., Hidaka, Y., Hara, K.: Removal of metal ions from aqueous solutions using carboxymethyl cellulose/sodium styrene sulfonate gels prepared by radiation grafting. Carbohyd. Polym. **157**, 335–343 (2017). https://doi.org/10.1016/j.carbpol.2016.09.049
99. Tu, H., Yu, Y., Chen, J., Shi, X., Zhou, J., Deng, H., Du, Y.: Highly cost-effective and high-strength hydrogels as dye adsorbents from natural polymers: chitosan and cellulose. Polym. Chem. **8**(19), 2913–2921 (2017). https://doi.org/10.1039/C7PY00223H
100. Tuan Zakaria, M.E., Jamari, S.S., Ling, Y.Y., Ghazali, S.: Synthesis of superabsorbent polymer via inverse suspension method: effect of carbon filler. IOP Conf. Ser. Mater. Sci. Eng. **204**(1) (2017). https://doi.org/10.1088/1757-899X/204/1/012013
101. Vundavalli, R., Vundavalli, S., Nakka, M., Rao, D.S.: Biodegradable nano-hydrogels in agricultural farming—alternative source for water resources. Procedia Mater. Sci. **10**(Cnt 2014), 548–554 (2015). https://doi.org/10.1016/j.mspro.2015.06.005
102. Wang, M., Guo, L., Sun, H.: Manufacture of Biomaterials (R. B. T.-E. of B. E. Narayan (ed.) pp. 116–134). Elsevier (2019). https://doi.org/10.1016/B978-0-12-801238-3.11027-X
103. Wang, T., Jones, J.D., Niyonshuti, I.I., Agrawal, S., Gundampati, R.K., Kumar, T.K.S., Quinn, K.P., Chen, J.: Biocompatible, injectable anionic hydrogels based on poly(oligo ethylene glycol monoacrylate-co-acrylic acid) for protein delivery. Adv. Ther. **2**(9), 1–9 (2019). https://doi.org/10.1002/adtp.201900092
104. Wei, D.W., Wei, H., Gauthier, A.C., Song, J., Jin, Y., Xiao, H.: Superhydrophobic modification of cellulose and cotton textiles: methodologies and applications. J. Bioresour. Bioprod. **5**(1), 1–15 (2020). https://doi.org/10.1016/j.jobab.2020.03.001
105. Wei, W., Qi, X., Li, J., Zhong, Y., Zuo, G., Pan, X., Su, T., Zhang, J., Dong, W.: Synthesis and characterization of a novel cationic hydrogel base on salecan-g-PMAPTAC. Int. J. Biol. Macromol. **101**, 474–480 (2017). https://doi.org/10.1016/j.ijbiomac.2017.03.106
106. Wu, R.L., Xu, S.M., Huang, X.J., Cao, L.Q., Feng, S., Wang, J.D.: Swelling behaviors of a new zwitterionic N-carboxymethyl-N, N-dimethyl-N-allylammonium/acrylic acid hydrogel. J. Polym. Res. **13**(1), 33–37 (2006). https://doi.org/10.1007/s10965-005-9002-1
107. Wu, X., Huang, W., Wu, W.-H., Xue, B., Xiang, D., Li, Y., Qin, M., Sun, F., Wang, W., Zhang, W.-B., Cao, Y.: Reversible hydrogels with tunable mechanical properties for optically controlling cell migration. Nano Res. **11**(10), 5556–5565 (2018). https://doi.org/10.1007/s12274-017-1890-y
108. Wu, X., Huang, X., Zhu, Y., Li, J., Hoffmann, M.R.: Synthesis and application of superabsorbent polymer microspheres for rapid concentration and quantification of microbial pathogens in ambient water. Separat. Purificat. Technol. **239**, 116540 (2020). https://doi.org/10.1016/j.seppur.2020.116540

109. Wu, X., Huang, X., Zhu, Y., Li, J., Hoffmann, M.R.: Synthesis and application of superabsorbent polymer microspheres for rapid concentration and quantification of microbial pathogens in ambient water. Sep. Purif. Technol. **239**, 116540 (2020). https://doi.org/10.1016/j.seppur.2020.116540
110. Xi, J., Zhang, P.: Application of super absorbent polymer in the research of water-retaining and slow-release fertilizer. IOP Conf. Ser. Earth Environ. Sci. **651**(4) (2021). https://doi.org/10.1088/1755-1315/651/4/042066
111. Xu, S., Wu, R., Huang, X., Cao, L., Wang, J.: Effect of the anionic-group/cationic-group ratio on the swelling behavior and controlled release of agrochemicals of the amphoteric, superabsorbent polymer poly(acrylic acid-co-diallyldimethylammonium chloride). J. Appl. Polym. Sci. **102**(2), 986–991 (2006). https://doi.org/10.1002/app.23990
112. Xu, X., Bai, B., Ding, C., Wang, H., Suo, Y.: Synthesis and properties of an ecofriendly superabsorbent composite by grafting the poly(acrylic acid) onto the surface of dopamine-coated sea buckthorn branches. Ind. Eng. Chem. Res. **54**(13), 3268–3278 (2015). https://doi.org/10.1021/acs.iecr.5b00092
113. Zhang, J., Xiao, H., Li, N., Ping, Q., Zhang, Y.: Synthesis and characterization of superabsorbent hydrogels based on hemicellulose. J. Appl. Polym. Sci. **132**(34) (2015). https://doi.org/10.1002/app.42441
114. Zohuriaan-Mehr, M.J., Omidian, H., Doroudiani, S., Kabiri, K.: Advances in non-hygienic applications of superabsorbent hydrogel materials. J. Mater. Sci. **45**(21), 5711–5735 (2010). https://doi.org/10.1007/s10853-010-4780-1
115. Zohuriaan, J., Kabiri, K.:. Superabsorbent Polymer Materials: A Review (2015) (June 2008)
116. Zou, W., Liu, X., Yu, L., Qiao, D., Chen, L., Liu, H., Zhang, N.: Synthesis and characterization of biodegradable starch-polyacrylamide graft copolymers using starches with different microstructures. J. Polym. Environ. **21**(2), 359–365 (2013). https://doi.org/10.1007/s10924-012-0473-y

Synthesis of Biopolymer Based Superabsorbent: An Eco-friendly Approach Towards Future Sustainability

Sweta Sinha

Abstract Acrylamide derivatives, methacrylic acid, and many other conventional synthetic vinyl monomers derived from petroleum are copolymerized to produce superabsorbents. However, most of these materials are expensive, environmentally hazardous, and inadequately biodegradable. In contrast to petroleum, biopolymers are derived from living things like bacteria and plants. The fundamental chemical building block of biopolymers is generated from renewable resources and is susceptible to environmental degradation. Biopolymers offer a number of remarkable benefits, including accessibility, economical fabrication, non-toxicity, biocompatibility, and biodegradability. As a result, biopolymers provide a fascinating and environmentally friendly alternative to traditional techniques. Biopolymers can easily form superabsorbents by chemical or physical crosslinking (including hydrogen bonding and ionic interactions), or a combination of the two, making it a versatile and promising method for fabricating superabsorbents (SABs). This research investigates strategies for novel synthetic methods to enhance the mechanical and physical properties of biopolymer based SABs from a bioengineering perspective. The paper starts with a brief introduction before entering into a thorough overview of synthesis and a comprehensive investigation of recent advancements of biopolymer based SABs. The primary properties of these materials, with a focus on swelling and mechanical properties, as well as the synthetic procedures utilized to make SABs from the biopolymers, are investigated. The unique properties of biopolymer based SABs, such as non-toxicity, biodegradability, biocompatibility, and eco-friendliness, make them excellent for use in a range of bioengineering sectors, including water treatment, medications, biomedicine, ecological, industrial, and food packaging. This study presents a basic overview of novel and cutting-edge bioengineering applications of biopolymer based SABs.

S. Sinha (✉)
Department of Chemistry, Amity Institute of Applied Sciences, Amity University Jharkhand, Ranchi 834002, Jharkhand, India
e-mail: sweta.sinha2203@gmail.com

S. Pradhan and S. Mohanty (eds.), *Bio-based Superabsorbents*, Engineering Materials,
https://doi.org/10.1007/978-981-99-3094-4_2

1 Introduction

Superabsorbents (SABs) are hydrophilic, three-dimensional, crosslinked materials that can absorb, swell, and retain an aqueous solution up to a thousand times their own weight [1]. SABs are primarily derived from water-soluble monomers to form a crosslink structure in the presence of a crosslinking agent and chemical initiator. The polymeric network of SABs can hold and absorb a lot of water or other aqueous fluids while preserving their structural dimensions [2]. These aqueous fluids can include blood, urine, electrolyte solutions, and much more. The absorbed fluid by SABs is difficult to remove even under pressure [3], unlike the conventional absorbent materials, namely tissue paper, cotton, polyurethane foams, etc. These properties are attributed to their enhanced entropy, water–polymer interaction, and osmotic force [4] induced by mobile ions.

Subsequent advances in the field of superabsorbent materials enabled even more impressive applications, such as agricultural, biomedical, and environmental applications. As a result, superabsorbent materials have become very popular in recent years due to their unique absorption capabilities [5]. Conventional SABs are primarily made of synthetic polymers derived from petroleum and composed of unsaturated vinyl monomers with hydrophilic groups, such as carboxylic acid in the case of acrylic acid (AA) and amides for acrylamide [6, 7]. These groups interact with water to form hydrogen bonds that capture it and increase its absorption capacity to 10–1000 times that of the dry superabsorbent [8, 9]. These materials conflict with the emergence of environmental consciousness because of their high production costs and lack of biodegradability. As a result, researchers are now turning to biopolymers, which are formed chemically from biological materials, as they are biodegradable, biocompatible, and nontoxic, even though their absorption capacity is lower and they have a shorter functional life than synthetic ones [10]. Biopolymers (Polysaccharides and Proteins) have received the most attention in terms of developing superabsorbents with significant swelling capacity. Although cellulose, starch, chitosan, and carrageenan are the most well-known for their ability to absorb water, many others have also been tested [11]. They are currently used in some specific applications, such as drug delivery or wound dressing in biomedicine, and as soil conditioners in agriculture [12, 13]. Due to the increasing demand for environmentally friendly products, biopolymers offer great potential as an alternative to synthetic superabsorbents.

The number of publications on polysaccharide-based SABs has increased significantly in the past 10 years, with many commercial applications in the hygiene sector. Industries are increasingly using natural polymers as superabsorbents, accounting for 80% of SAB production worldwide. This review will focus on the types and classification of SABs, their structures and properties, methods of preparation and characterization, and potential applications.

2 Definition and Classification

The International Union of Pure and Applied Chemistry (IUPAC, iupac.org) defined a SAb in 2004 as "a polymer that can absorb and retain extremely large amounts of a liquid relative to its own mass." It has an absorption capacity of 1,000 times its own weight for water or an organic liquid [14]. They are classified as "superabsorbent hydrogels" when they absorb water at a rate that is greater than 20 times their initial weight [15]. SABs are frequently composed of ionic monomers, and their low crosslinking density allows them to have a higher fluid absorption capacity.

2.1 Classification of Superabsorbent

Superabsorbents can be classified based on morphology, source, crosslinking mechanism, electrical charges, and responsiveness to stimuli, polymeric composition, or polymer chain configuration. An overview of these classifications is shown in Fig. 1.

2.1.1 Classification Based on Sources

Superabsorbents are composed of polymers that can be natural, synthetic, or semi-synthetic in nature [13]. Petrochemical byproducts known as synthetic monomers are used to create synthetic superabsorbents, such as polyacrylic acid, polyethylene glycol, and polyacrylamide [8–10]. These polymers are inexpensive and have a high degree of mechanical strength, but they are not biocompatible and degrade inadequately [13]. Superabsorbents can also be synthesised from natural polymers like polysaccharides or proteins, which are derived from natural resources and have

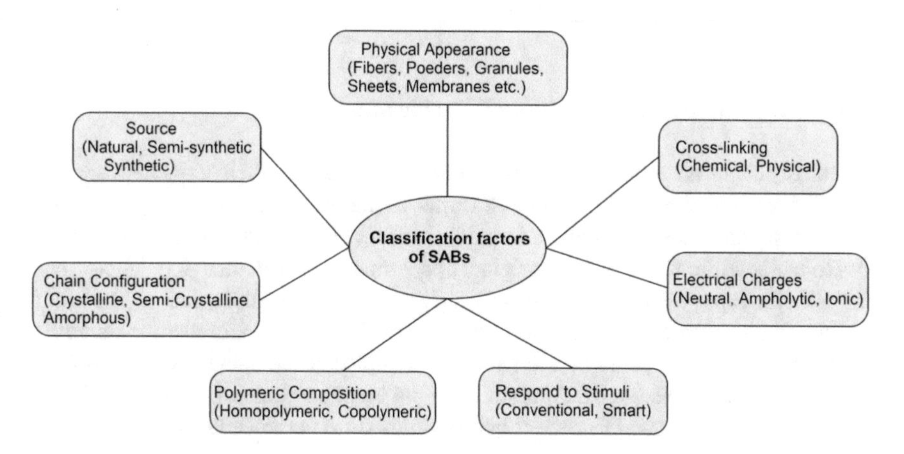

Fig. 1 Classification of SABs based on different factors

high biocompatibility, biodegradability, and nontoxicity. The third type of superabsorbent is semisynthetic, which blends natural and synthetic polymers and is formed by grafting a synthetic polymer to a natural backbone. However, these copolymers are only partially biobased and biodegradable, which limits their applications [13].

2.1.2 Classification Based on Crosslinking

Crosslinking superabsorbent is the process of forming a polymeric network by forming covalent bonds with strong energies greater than 100 kJ/mol. The primary chemical process is free-radical polymerization [8], but a polymer backbone and a crosslinker are required for all chemical crosslinking. In the presence of chemical reagents used as crosslinkers, polysaccharides like chitosan, alginate, starch, or cellulose can form chemical superabsorbents [15], which are reversible or nonpermanent due to noncovalent and weak bonds with energies up to 40 kJ/mol, electrostatic interaction, hydrogen bonding, and hydrophobic interaction [16]. This kind of crosslinking lacks toxic crosslinking agents, allowing it to be extensively used for applications in the pharmaceutical and biomedical fields [8], such as when starch and carboxymethylcellulose are combined [15].

2.1.3 Classification Based on Electrical Charges

Superabsorbents are divided into four categories based on the presence of electrical charges on the side chains or backbone of the superabsorbent. Neutral polysaccharides such as dextran or agarose are examples of nonionic or neutral superabsorbents, while anionic or cationic species form ionic superabsorbencies. Synthetic monomers with anionic and cationic properties are AA, or aminoethyl methacrylate respectively, which are biocompatible and can be created using synthetic monomers such as maleic anhydride or N-vinyl succinimide [15].

2.1.4 Physical Appearance-Based Classification

Superabsorbents are classified morphologically based on their expected applications. They are available in a variety of forms, including fibres, powders, emulsions, granules, membranes, and sheets. Superabsorbents should have enough strength to absorb water without losing their original appearance and structure [13]. Microfibers of synthetic superabsorbent based on polyacrylate, for example, are used as self-healing coatings on carbon steel to prevent corrosion [17]. Superabsorbents with a granular appearance made of starch and ethylcellulose have also been used as coatings to encase fertilisers [18]. Another example is in the biomedical sector, where a SABs-based sheet is used to collect bodily fluids from patients during surgery [9].

2.1.5 Classification Based on Response to Stimuli

Superabsorbents are referred to as "smart" or "environmentally responsive" when their three-dimensional conformation varies in response to changing conditions. They can respond to stimuli such as temperature, pH, ionic strength, light, pressure, and electric and magnetic fields and are used in drug delivery. pH-sensitive alginate-based superabsorbents, for instance, have been studied [19].

2.1.6 Classification of Polymeric Composition

Superabsorbents can be crosslinked networks of polymers composed of a single polymer species. Homopolymeric superabsorbents are made up of two or more different polymers with one or more hydrophilic groups. Carboxymethyl cellulose and carboxymethyl chitosan were combined to form a copolymeric material for metal absorption. When two different crosslinked polymers interpenetrate one another while being crosslinked by a crosslinking agent, the result is a network of interpenetrating polymers (IPN). Alginate and hydrophobically altered ethyl hydroxyl ether cellulose are two examples of this kind of material [8]. Semi-IPNs exist when only one polymer is crosslinked and the other is not, such as linear carboxymethylchitosan and crosslinked synthetic polymers. Recent studies have focused on a superabsorbent semi-IPN with a starch backbone and grafted synthetic monomers containing clinoptilolite and polyvinyl alcohol.

2.1.7 Classification Based on Chain Configuration

The latest superabsorbent classification is based on chain structure or configuration. Assimicrystalline materials, blends of the two previous ones, as well as amorphous and crystalline superabsorbents (orderly arrangements), have been reported [20]. Chitin can have a semicrystalline morphology with acetyl groups distributed heterogeneously along the polymer chains [21], while starch can have a crystalline morphology with amylose and amylopectin.

2.2 Biopolymer as Superabsorbent

A polymer derived from biological materials is known as a biopolymer. Biopolymers are increasingly being explored for their use as superabsorbents due to their intriguing properties and diverse applications. The two biopolymers most prevalent in nature are cellulose and chitin, which are derived from terrestrial plants and animals, respectively. They have been investigated for a variety of applications in the industrial, biomedical, pharmaceutical, drug-delivery, tissue engineering, medications, tablets, adhesives, paper, food, cotton, and rayon sectors, including because of their abundance, high tolerance, high thermal stability, non-toxicity, antibacterial activity, antifungal activity, biocompatibility, and other distinctive qualities like strong adsorption capacities and ease of functionalization [22]. Polysaccharides are found in all living things, including seaweed and microorganisms, and can be branched or linear. Alginate, chitin and chitosan, starch, carrageenan, agar and agarose, pectin, hyaluronic acid, dextran, cellulose, and gums are the absorbent biopolymers that are most frequently discussed in the literature [11]. Chemical modifications can improve their physicochemical properties and give them new advantageous properties, so their derivatives are being studied for superabsorbent applications.

3 Characterization of Biopolymer Based SABs

Characterization of biopolymers is essential for designing new materials, evaluating competing products, and improving product performance. It includes information on the molecular weight distribution, molecular structure, shape, thermal properties, mechanical properties, and any modifications made (Fig. 2). The physical characteristics of biopolymer membranes and films—such as swelling, degree of degradation or erosion, and porosity measurements—are described below. The explanation of polymer biological properties, such as cytotoxicity and antimicrobial research.

3.1 Physical Characteristics

3.1.1 Swelling Capacity

Superabsorbents are materials that have been studied for their ability to absorb and retain water or physical and biological solutions, such as urine, electrolyte solution, and blood [6]. Furthermore, they have higher absorption than traditional absorbent materials, as hydrogels can absorb up to a thousand times their own dry weight in water without dissolving [5, 6]. This incredible property is very appealing for a variety of applications. However, various factors, including the type of monomers, the density of the crosslinks, and other stimuli, affect the swelling ability [11]. Later in this review, the impact of these parameters on the swelling capacity is discussed. Superabsorbents can swell through a variety of processes and mechanisms [23]. The general mechanism of water absorption is described below. Water molecules enter the polymer network first and hydrate the most polar hydrophilic groups (like carboxylates). Other hydrophilic groups such as alcohol (OH), amides (CONH, CONH2), or sulfonic acid (SO3H) may also be present [8]. When these groups form hydrogen bonds with water, they form "primarybound water." After hydration, the polymer network expands, and "secondary bound water" results from interactions between hydrophobic groups and water. Total bound water is made up of both primary and secondary bound water. In polysaccharides, hydroxyl groups (–OH) allow hydrogen

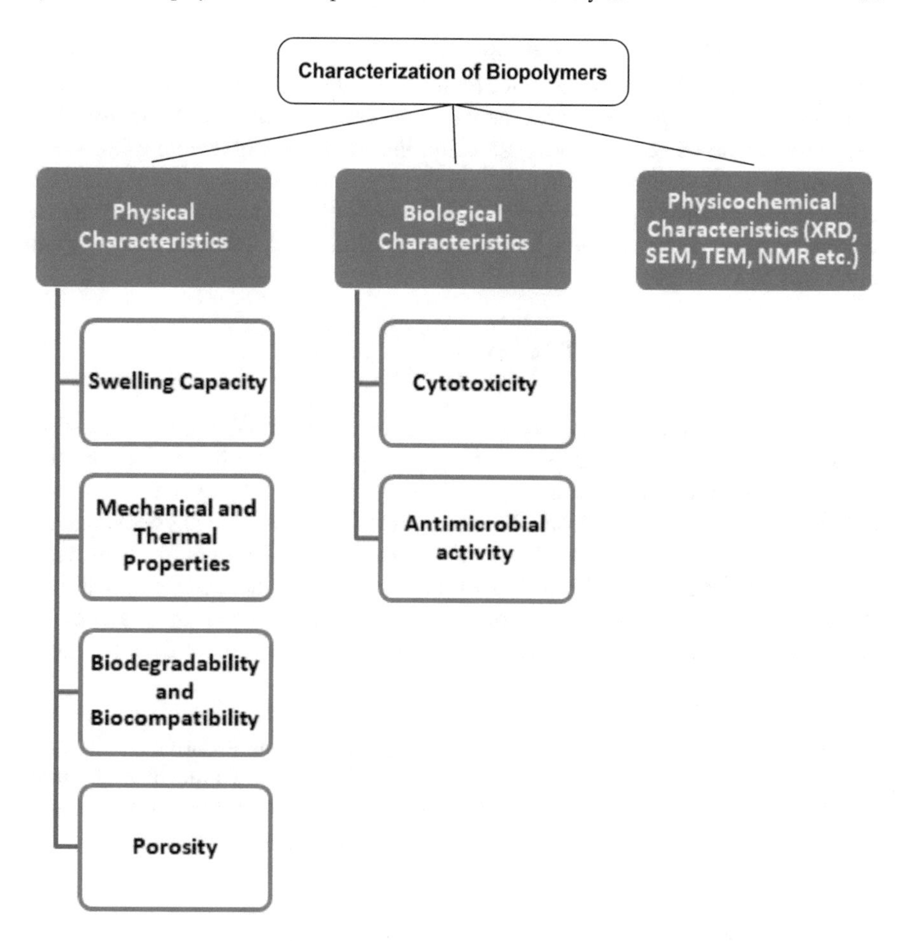

Fig. 2 Different characteristics of biopolymer

bonds to form, whereas –CH groups impart a hydrophobic character [21]. The spaces between network chains and pores will then be filled with additional water up to the equilibrium swelling level. It's referred to as "free water" or "bulk water." This phenomenon is a result of the network's osmotic force, which tends towards infinite dilution and opposes physical crosslinking, causing the network to become elastic by retraction [6, 8, 23].

$$\text{Swelling Degree}\,(\%) = \frac{W_t - W_0}{W_0} \times 100\% \tag{1}$$

Or,

$$\text{Swelling Ratio}\,(SR) = \frac{W_t - W_0}{W_0} \tag{2}$$

where W_0 is the initial weight of the dry superabsorbent and W_t is the weight of the swollen superabsorbent at the specified temperature and time. The most common unit of measurement is g/g of superabsorbent [1, 24]. The terms of Equilibrium Swelling (S_{eq}), Equilibrium Swelling Ratios (SR), Equilibrium Degree of Swelling (EDS) and Percentage swelling are used when the weight of the swollen superabsorbent is taken at swelling equilibrium ($W_t = W_{eq}$) [8, 24]. "Water retention capacity" is the volume of water released after reaching swelling equilibrium. It is defined by Eq. (3) and expressed as a percentage:

$$WR(\%) = \frac{W_t - W_0}{W_i - W_0} \times 100 \quad (3)$$

where W_i is the initial weight of the swollen superabsorbent, W_0 is the weight of the dry superabsorbent, and W_t is the weight of the superabsorbent heated for a specific period of time and temperature. Superabsorbent made of biopolymers has a high capacity to hold water. A chitosan derivative graft called AA SAP has demonstrated good retention ability. Indeed, after reaching swelling equilibrium, it could retain 71% of its water for 24 h. It took 120 h to release all of the water that had been absorbed. This is possible because of the numerous interactions (hydrogen bonds and Van der Waals forces) between water and the polymer, as well as the small network configuration [25].

The swelling degree of biopolymers can be used to predict material behavior after application. Because they provide better barrier properties, low swelling qualities are required for food packaging applications [26]. Swelling is a key feature in rehydration and exudates absorption for wound-healing applications. Moisture must be provided in sufficient quantities to prevent dehydration, bacterial growth, and infection [27].

Swelling Properties

Superabsorbents' swelling capacities are affected by various variables, such as crosslinking, ionic strength, and composition [5, 28]. Some "smart" superabsorbents can change their swelling characteristics in response to environmental stimuli, such as temperature, pH, light, or pressure [13].

Different parameters influence swelling

(a) **Physical Stimuli**

Superabsorbents are materials that change volume in response to temperature changes and exhibit behaviours such as swelling or shrinking. At the lower critical solution temperature (LCST) [29], they are hydrophilic and can absorb liquids below LCST, while at higher temperatures, they condense and become insoluble. Superabsorbents can also exist and remain constant below UCST. Polysaccharides, such as cellulose and dextran, can develop into smart superabsorbents with thermoresponsive behaviour [8]. A superabsorbent grafted onto kappa-carrageenan exhibits a significant increase in water absorption from 50 to 80 °C, with a maximum absorbency of 789 g/g, followed by a decrease above 80 °C. A thermosensitive pectin-based superabsorbent exhibits a similar trend, increasing its capacity to swell with water from 376 g/g to a maximum of 484 g/g by varying the temperature from 8 to 37 °C. This increase was attributed to temperature-related factors, such as the flexibility of polymer chains or the diffusion of water molecules in the network [4].

Superabsorbents with thermosensitivity and light-sensitive or photoresponsive properties have limited applications in biomedical fields due to ultraviolet radiation's inability to penetrate tissues. Ionic polymers are used to develop superabsorbents that are electrically sensitive due to the formation of a concentration gradient that modifies osmotic pressure and causes shrinking or swelling behaviour.

(b) **The Effect of Chemical pH Stimuli**

The swelling and absorption abilities of superabsorbents can be influenced by pH. The size of these materials can change significantly as a result of even small changes in an aqueous solution's pH. These alterations are primarily the result of the polymeric network's ionic groups, which, depending on the pH of the aqueous medium, can be protonated or deprotonated (ionised) and are highly hydrophilic. Carboxylic acid (–COOH) and sulfonic acid (–SO3H), for example, are protonated below their pKa (pH pKa), whereas asamines are protonated when the solution's pH is higher than the pka (pH > pKa). M When these groups are ionised (or deprotonated), this leads to increased swelling because of osmotic force.

When acidic groups are protonated, the superabsorbent swells significantly due to an increase in anisotropy for water. This phenomenon is explained by strong hydrogen bonds between acidic groups and an acidic medium (defined as pH pKa). Acidic groups become charged when the pH is higher, which weakens hydrogen bonds and increases electrostatic repulsions between anionic groups. Counterions, particularly sodium cations (Na+), can have a screening effect in highly basic solutions (pH > 9).

This mechanism was discovered while studying the effects of pH on swelling capacity in a pH-sensitive superabsorbent made of pectin. The protonation of the carboxylic acid groups in pectin causes the swelling value in the acidic buffer to rise to 22.4 g/g, but in the base solution at pH 7.4, these groups are charged (–COO), and electrostatic repulsions provide a maximum swelling of 88.3 g/g [4]. A study on a semi-synthetic superabsorbent with AA grafted onto a chitosan backbone showed that most groups are protonated at pH 4–6, resulting in swelling of less than 350 g/g. At pH values between 6 and 8, electrostatic repulsions increase to a maximum of 530 g/g, and carboxylic acid groups become ionised. The screening effect of counterions causes a decrease in the water absorption capacity at pH levels lower than 2 or higher than 12 [25].

(c) **Effect of Ionic Strength**

Superabsorbent swelling and absorption properties are influenced by the ionic strength of swelling solutions [5], with a saline reducing the swelling ratio. For example, a pectin-based superabsorbent has a maximum absorption of 396 g/

g with distilled water, but a swelling ratio of 74 and 47 g/g in 0.05 and 0.5 M of NaCl solution, respectively [4]. This is similar to a hemicellulose-based superabent that exhibits a swelling of 1128 g/g [30]. Decreasing the swelling of superabsorbents can be achieved by increasing the external ionic concentration due to cations reducing theosmotic pressure and electrostatic repulsions between anionic groups [8].

The charge screening effect permits additional ionic crosslinking in the polymeric network, preventing expansion and lowering the swelling capacity. The size and valency of cations are related to swelling, with divalent and trivalent ions forming stronger complexes with carboxylate ions. NaCl and KCl solutions have a higher swelling capacity than $MgCl_2$, $CaCl_2$, and $AlCl_3$ saline solutions [11]. This effect was seen in $CaCl_2$ and NaCl solutions using a superabsorbent hydrogel based on cellulose. The higher ionic strength of $CaCl_2$ solution caused the swelling ratio to decline more quickly than that of NaCl solution. The swelling ratio was 1000 g/g without salts and decreased to 225 g/g at 0.1 M NaCl [31].

(d) **Effect of concentration (Monomer and Initiator)**
Studies have shown that the concentration of monomer units and initiators affects the grafted superabsorbent' absorption capacity. Water absorption usually rises substantially with initiator concentration until a maximum and then sharply declines. For example, incorporating AA increased the water absorption of a chitosan derivative superabsorbent from 241 to 429 g/g when the amount of the initiator (ammonium persulfate) used increased from 0.5 to 2.0% [30]. The lack of active centres in polymeric networks can lead to lower water absorption. As the initiator quantity increases, it is possible to improve water absorption, with 2.0% of initiator being the maximum. However, excessive initiator content can lead to shorter chain lengths or compact networks, reducing the capacity to absorb water [30]. An increased concentration of grafted synthetic monomers has a similar effect on their capacity to absorb water as an increased concentration of the initiator. This is demonstrated by a hemicellulose-based superabsorbent with grafted AA and amide. Water absorption increased with increasing AA content, reaching a maximum of 997 g/g (for 15 g of AA). Absorption began to decrease after this point, due to excessive crosslinking.

(e) **Effect of Neutralization**
Neutralization of anionic charges on biopolymer grafted SABs with alkaline solutions affects their swelling capacities [32]. Swelling and absorption capacities are related to carboxylate electrostatic repulsions and anisotropy to water. When AA monomers are neutralised by NaOH [33], they become charged, allowing for an increase in swelling due to repulsive forces. The neutralisation degree of hemicellulose grafted with AA increases from 65 to 75% and reaches a maximum of 1100 g/g, increasing its swelling capacity. Water absorption drops to less than 800 g/g as the neutralisation degree is further increased [30].

(f) **Effect of Crosslinker and Crosslinking Density**
Cross-linking density affects the mechanical properties and swelling ratio of a three-dimensional network, determined by the type and concentration of

crosslinking agent used [8]. Superabsorbent swelling is generally inversely proportional to crosslinking density and, thus, elastic modulus [28]. In other words, increasing the crosslinking density reduces the gel's capacity to swell while increasing its resistance [16]. More specifically, a higher crosslinking density tends to result from a higher crosslinker concentration. As a result, the spacing between chains in the polymeric network is reduced, and the superabsorbent's structure is closer and more stable. The network's elasticity decreases, making it more difficult for water molecules to enter the three-dimensional structure, resulting in low swelling and absorption capacity [32, 34]. Regardless, when insucient crosslinking occurs, superabsorbent materials can partially dissolve in aqueous solutions and continue to lose absorption capacity [13]. This phenomenon can be illustrated by a hemicellulose-based superabsorbent hydrogel. With this substance, the rate of crosslinker (N,N′-methylenebisacrylamide) to monomer increased rapidly to 0.03% (w/w), which resulted in an increase in water absorption. When the crosslinker content was significantly high, the swelling ratio declined to under 600 g/g after reaching a maximum of 1128 g/g. Saline solution absorption showed a similar effect, but with a maximum absorption of 132 g/g [30]. Equation (4) establishes a correlation between swelling and crosslinker concentration:

$$\mathrm{Swelling} = \mathrm{K} \cdot \mathrm{C}_\mathrm{C}^{-\mathrm{n}} \tag{4}$$

where $\mathrm{C_C}$ is the concentration of the crosslinking agent and k and n are specific constants for each superabsorbent.

3.1.2 Mechanical and Thermal Properties

Superabsorbents' mechanical strength is a key property for biomedical and pharmaceutical applications such as tissue engineering, drug delivery, and wound dressing. In fact, this characteristic enables superabsorbents to maintain their physical form. According to studies, it can either be increased by enhancing the crosslinking density to produce stronger superabsorbents or, on the other hand, decreased by heating [8]. The mechanical properties of many superabsorbents, including those based on polysaccharides, are, however, weak [8]. They are indeed brittle, have limited extensibility and recoverability, and lack an effective energy dissipation mechanism [11]. Hooke's law (3), which is usually used to define elastic solids, can also be used to observe the mechanical behaviour of superabsorbents.

By using the Young's modulus as the proportionality constant, it shows that a solid's strain is inversely proportional to the applied stress [5, 25].

$$\sigma = \mathrm{E} \cdot \acute{\epsilon} \tag{5}$$

In Eq. (5), σ is the applied stress (Pa), "is the strain or the deformation," $\acute{\varepsilon}$ and "E" is the Young's modulus (Pa).

Temperature-responsive superabsorbents do exist, though their thermal properties are less well understood. They are useful in biomedical applications because they can change shape as the temperature changes. Temperature-responsive superabsorbent hydrogels based on cellulose, chitosan, or dextran are feasible [8]. The mechanical and thermal characteristics of superabsorbents must be enhanced to expand the range of applications for them. Diverse strategies have been used to address this issue, such as the incorporation of particular crosslinkers or the synthesis of double networks, triblock copolymers, and nanocomposites. The mechanical and thermal characteristics of superabsorbents are then investigated and characterised using differential scanning calorimetry (DSC) and dynamic mechanical analysis [5].

3.1.3 Biocompatibility and Biodegradability

Unlike synthetic superabsorbents, most natural superabsorbents are biodegradable. This signifies that enzymes and microorganisms can break the bonds between and within polymeric structures [8]. Research on biopolymer-based superabsorbents is influenced by their ability to degrade, which may be able to address environmental issues in the biomedical and agricultural fields [11]. Biocompatibility is highly desired in tissue engineering to ensure positive interactions with the human body and prevent damage to the connected tissues. Cell culture techniques or cytotoxicity tests can be used to evaluate the tissue biocompatibility of SABs.

Biopolymers are becoming more and more biocompatible, allowing them to be incorporated into a host with the proper response for a given application. SABs can be used to create antibacterial superabsorbents and reduce toxicity to mammalian cells. Biocompatibility is important for tissue engineering, and cell culture techniques or cytotoxicity tests are used to evaluate their tissue biocompatibility [35].

3.1.4 Porosity Measurements

The ability of polymer film membranes to absorb water can be assessed physically by measuring the porosity of the membranes. The method involves A small section of uniformly sized, pre-weighed polymer film was immersed into the alcohol solution. After a short while, the film fragments are taken out, and the ultimate weight is determined. Equation 6 shows a mathematical formula for calculating the porosity of a polymer film membrane.

$$\mathbf{Porosity(P)} = \mathbf{W}_2 - \mathbf{W}_1/\boldsymbol{\rho}\,\mathbf{V1} \tag{6}$$

where W_1 is the initial weight of the dry film and W_2 is final weight of the film after immersion in the alcohol solution. V_1, the volume of the solution taken before submerging the film, ρ measures the density of the solution.

3.2 Biological Characterization

3.2.1 Cytotoxicity

According to international recommendations to decrease in vivo testing, the majority of tests are now performed in vitro. The first three biocompatibility tests can be performed using cell cultures, either primary or cell lines, to assess the effect induced by the biomaterial's exposure to the cells. All biomaterials are subjected to the cytotoxicity test, which uses the cytotoxicity endpoints of neutral red uptake and BALB/c 3T3 mouse fibroblasts or healthy human epidermal keratinocytes. It is possible to use additional essential dyes, such as 3-(4,5-dimethyl-2-thiazolyl)-2,5-diphenyl-2H-tetrazolium bromide (MTT) or 3-(4,5-dimethyl-2-thiazolyl)-5-(3-carboximetoxiphenyl)-2-(4-sulfophenyl)-2H-tetrazolium bromide (MTT) [36]. Sensitization is a term used to describe the toxicological endpoint that is connected to substances that have the ability to inherently cause skin sensitivity. This undesirable effect results from an overreaction of the adaptive immune system. As a result of subsequent interactions, sensitization was produced [37]. Sensitization was initially induced upon contact with allergy symptoms.

3.2.2 Antimicrobial Activity

The biostatic or biocidal microbiological characteristics of films in solid or liquid media can be identified in terms of their antimicrobial activity. Antimicrobial activity can be evaluated using macrodilution and microdilution in broth methods specified by the Clinical and Laboratory Standards Institute. Macrodilution is used, which involves using larger volumes and more samples. In both cases, the microorganism was added in varying amounts to the tubes (macrodilution) or 96 microplate wells (microdilution), and the films were added in proportion. Both positive and negative controls used to be present. Depending on the investigated microorganism, the final concentration was established as the lowest concentration that, after 24 h of incubation at 37 °C, produced no detectable bacterial development. Pure medium served as the control. Placing a 100-mL solution from the test tubes onto agar plates devoid of obvious bacterial growth and incubating for an additional 24 h at 37 °C allowed us to determine the biostability level. If there is growth, the films are regarded as biostatic; however, if there is no growth, they are also regarded as biocides [38].

4 Preparation and Synthesis of Biopolymer-Based Superabsorbents

4.1 Materials

Sanyo Chemical was the first company to market a SAB in 1978, but commercial manufacturers rarely discuss the synthesis of SABs from biopolymers. After being grafted onto starch and polymerized from acrylamide, it was neutralized, dried, and powdered [39]. SABs are typically made with monomers or polymers, an initiator, and a crosslinking agent [20]. The IUPAC defines polymerization as the process that transforms a monomer or mixture of monomers into a polymer, while the monomer molecule is defined as "a molecule that can undergo polymerization, thereby contributing constitutional units to the essential structure of a macromolecule" [40]. The definitions of these two terms are "reaction involving sites or groups on existing macromolecules" and "a molecule of high relative molecular mass, the structure of which essentially consists of the multiple repetitions of units derived, actually or conceptually, from molecules of low relative molecular mass." We discuss polymerization when monomers are grafted onto a crosslinked polymer network. Crosslinking is an essential step in the preparation of macromolecules such as SABs By virtue of their functional groups, biopolymer-based SABs can be crosslinked with one another as well as with other polymers and molecules. N,N′-methylenebisacrylamide, glyceraldehyde, formaldehyde, epichlorohydrin, and adipicdihydrazide are all examples of crosslinkers [15]. The initiator is a chemical that stimulates the grafting-induced polymerization of monomers on polymeric backbones. Grafting may necessitate the use of synthetic monomers such as AA or acrylamide. For example, AA monomers were grafted onto kappa-carrageenan with ammonium persulfate as an initiator [32]. The use of aqueous solutions as diluents and washing solutions for removing impurities are also effective [20]. A neutralizing agent, such as sodium hydroxide, will also need to be added if the neutralization step needs to be performed during the synthesis of superabsorbent [23].

4.2 Crosslinking Methods

Biopolymers may be physically or chemically crosslinked to create SABs (Fig. 3). Chemical crosslinking creates irreversible and stable superabsorbents that can withstand changes in the external environment by forming strong covalent bonds between polymer chains. This type of SAB has good mechanical strength and a longer degradation time, but the toxicity of some crosslinkers can be challenging for some applications [8]. Physical crosslinking, in contrast, enables reversible SABs in which polymers are linked by secondary forces such as ionic or hydrophobic interactions and hydrogen bonding. This type of crosslinking does not require a crosslinker, but the stability of SAB is dependent on external or physical variations [15]. There are

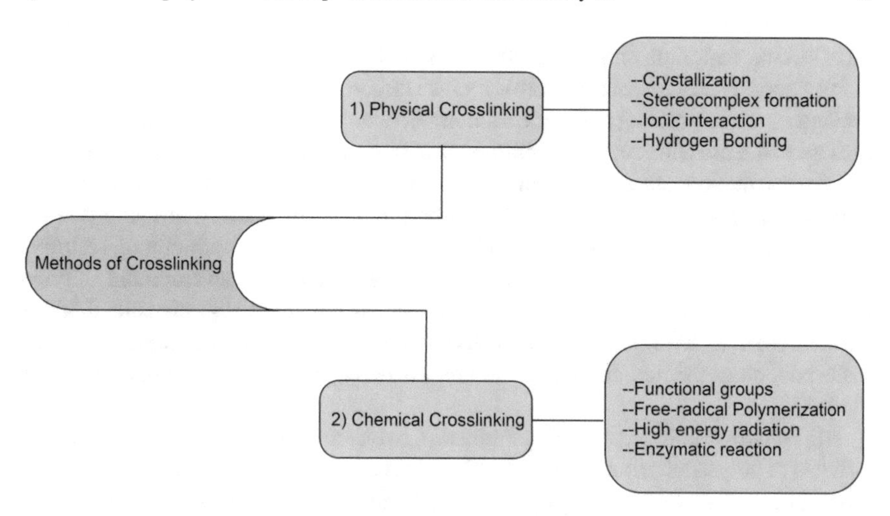

Fig. 3 Method of crosslinking for the formation of SABs

numerous chemical and physical crosslinking techniques that have been reported to develop SABs [6].

4.2.1 Chemical Crosslinking

Polysaccharide-based SABs can be formed using a variety of chemical crosslinking techniques. The reported mechanisms include crosslinking by functional groups, free-radical polymerization, high-energy radiation, and enzymatic reactions [8]. In order to produce SABs, a crosslinker can chemically crosslink the functional and hydrophilic groups presented in biopolymers, such as carboxylic acids (–OOH), amines (–NH2), and hydroxyl groups (–OH). These groups are used in three different mechanisms: schi-basis, Michael addition, and condensation reactions. Amine groups in polysaccharides can interact with aldehydes or dihydrazides as crosslinkers to create schi-bases, which can then transform into SABs. The most widely used crosslinking agents are glutaraldehyde and adipic acid dihydrazide. They function to produce an imine group at high temperatures, low pH, and the addition of methanol (Sch-base). Hydrogel can be formed when amines of modified hyaluronic acid react with aldehyde groups of polyethylene glycol propionaldehyde. SAHs are also formed when an electrophile, such as vinyl or acrylate groups on another polymer, reacts with a nucleophile function of a polymer, such as amine or thiol groups, via the Michael addition. Michael addition is carried out using polysaccharides like chitosan, dextran, and hyaluronic acid. Finally, condensation reactions between hydroxyl or amine groups and carboxylic acids or their derivatives can lead to the establishment of ester or amide bonds.

These reactions necessitate slightly acidic water and room temperature. N,N-(3-dimethylaminopropyl)-N-ethyl carbodiimide and N-hydroxysuccinimide are the crosslinking materials employed [8]. Another chemical crosslinking technique that is frequently used to create superabsorbents is free-radical polymerization [13]. This method involves grafting synthetic monomers onto polysaccharide backbones in the presence of a crosslinker, most commonly N,N′-methylenebisacrylamide [11].

Polymerization starts with initiators, which can produce free radicals from polysaccharides. These radicals can also be produced by heating or using a redox system. The radicals will then be able to react with synthetic monomers, resulting in chain propagation. In addition, the crosslinker enables the development of polymeric networks. The most widely used polysaccharides are hyaluronic acid, dextran, and chitosan, as well as acrylic derivatives like vinyl monomers. Emulsions, bulk materials, suspensions, and solutions can all undergo free-radical polymerization [1, 8, 11].

High-energy radiation such as gamma rays or electron beams can also produce free radicals in polysaccharide chains. However, this method takes place at room temperature and doesn't require a crosslinker. Radiation is absorbed by water molecules during crosslinking in an aqueous solution, creating radicals that activate the polysaccharide backbone. Although polysaccharide-based SABs are typically crosslinked in an aqueous solution, the irradiation can also occur in a concentrated solution or in a solid state. SAB was created by using high-energy radiation to crosslink Arabic gum, carboxymethyl cellulose, and dextran in a solid state [8, 15].

4.2.2 Physical Crosslinking

Crystallization, stereocomplex formation, ionic interaction, hydrogen bonding, maturation, and hydrophobic interactions are the methods used for physical crosslinking of polysaccharide-based SABs [8].

Crosslinking by Crystallisation

It is primarily used for thermosensitive SABs that change shape with temperature. Double helices and crystallites both develop at room temperature and serve as crosslinking sites to build the polymeric network. The freezing–thawing procedure is typically repeated several times for the crystallisation method. SABs have been reported for chitosan, cellulose, and dextran [8].

Crosslinking by Ionic Interaction

SABs can also be formed by ionic interactions between polyelectrolytes, which are polymers with opposing charges. A multivalent ion or a charged polymer can also interact with a molecule with an opposing charge. All of these polymers can be crosslinked at physiological pH and room temperature. For example, alginate can be crosslinked with calcium ions, chitosan with glycerol-phosphate disodium salt, and carrageenan with potassium ions.

Crosslinking by Forming Stereocomplexes

The formation of a stereocomplex is another physical crosslinking technique. It is based on interactions between polymers with various chiralities that come together to form a stereocomplex. This approach makes it simple to synthesise SABs by blending the solutions after each polymer has been dissolved in an aqueous solution at room temperature. It has been reported that SAB hydrogel can be formed using dextran and grafted lactic acid oligomers. However, this method is constrained by the small number of polymers with stereocomplex properties [8, 15].

Crosslinking by H-Bonding

The presence of functional groups in polysaccharides allows for hydrogen bonding as well. Hydrogen bonding can crosslink two polymers, a polymer and an acid, or a polyfunctional monomer. On cellulose, alginate, and chitosan, functional groups such as hydroxyl, carboxyl, and amine groups readily form hydrogen bonds [8].

Maturation is the process by which natural polymers crosslink to form a crosslinked structure, also known as heat-induced aggregation. Interactions between hydrophobic polymer backbones building blocks formed in an aqueous solution are used in the final physical crosslinking technique. These interactions could be brought on by dextran or chitosan [28].

5 Applications

SABs have a wide range of applications. The most common use is in disposable personal hygiene items like feminine napkins, adult incontinence products, and baby diapers [11, 41, 42]. SABs are important in agriculture and forestry because they reduce water consumption, reduce plant death, improve fertilizer retention, and improve plant growth rates [43, 44]. They also have an impact on the water's permeability, density, structure, texture, evaporation, and infiltration rates. Irrespective of its ability to improve soil conditions, it must overcome the challenges posed by saline-containing water and soils.

SABs are employed in the biomedical and pharmaceutical industries because of their ideal characteristics, including their soft and rubbery consistency, high water content, and low interfacial tension with water or biological fluids. They are hydrophilic and inert, which makes them perfect for a variety of bio-related applications, including controlled drug delivery systems, biosensors, soft contact lenses, cartilage reconstruction, and blood-contacting biomaterials [45]. SABs have been used in a variety of applications, including wire-and-cable water blocking, filtration applications, spill control, hot and cold therapy packs, composites and laminates, medical waste solidification, mortuary pads, motionless waterbeds, candles, waste stabilization, environmental remediation, fragrance carriers, wound design, fire protection, surgical pads, controlled release, grow-in-water toys, and many others [46, 47]. Modifications to SABs' properties have been made due to their increased usage. Such modifications have allowed SABs to be used in many more applications and industries (Fig. 4), such as cosmetic products, hygiene products, food processing and packaging, paint additives, adhesives, etc., [11, 43–49].

Fig. 4 Spectrum of applications of biopolymer based SABs

6 Conclusions

Superabsorbents have attracted attention in recent years due to their ability to absorb and retain large amounts of fluid. The majority of commercially available superabsorbents are composed primarily of synthetic polymers. The most appropriate and widely used classification is based on the origin of the crosslinked polymers. Their absorption capacity is extremely high, making them suitable for a variety of applications; however, new environmental concerns about toxic waste are prompting manufacturers to turn to more natural materials. Because they have a high absorption capacity and are biodegradable and biocompatible, natural polymers are being used to create superabsorbents to meet these demands. They have a lower absorption capacity than their synthetic counterparts, but they are still useful in a variety of applications, including agriculture and the biomedical field. Biopolymer-based superabsorbents have been investigated for potential applications in a variety of fields. Characterization techniques have been used to evaluate the properties and structures of these materials. It is also possible to graft synthetic monomers or polymers onto these biosourced polymers in order to combine their properties for specific applications. According to the data in this review, the market for superabsorbent materials will probably keep expanding. Future production of polysaccharide-based superabsorbents may increase due to the low cost and abundance of natural polymers, which also have intriguing properties for promising applications. Toxic synthetics should be restricted or outlawed in order to increase the market for natural superabsorbents. The emergence of new regulations would encourage the development of methods for manufacturing natural superabsorbents, and consumers should adopt new products based on biopolymer SABs to offset their large-scale production costs. In summary, the superabsorbent market is expanding and has the potential to become a sustainable industry by modifying the properties of bioresources polymers.

References

1. Snoeck, D.: Self-Healing and Microstructure of Cementitious Materials with Microfibres and Superabsorbent Polymers. Ghent University, Ghent, Belgium (2015)
2. Jindal, N., Khattar, J.S.: Microbial polysaccharides in food industry. In: Biopolymers for Food Design. Academic Press, Cambridge, pp. 95–123 (2018)
3. Pourjavadi, A., Barzegar, S.H., Mahdavinia, G.R.: MBA-crosslinked Na-Alg/CMC as a smart full-polysaccharide superabsorbent hydrogels. Carbohydr. Polym. **66**, 386 (2006)
4. Pourjavadi, A., Barzegar, S.: Synthesis and evaluation of pH and thermosensitive pectin-based superabsorbent hydrogel for oral drug delivery systems. Starch Stärke **61**, 161–172 (2009)
5. Guilherme, M.R., Aouada, F.A., Fajardo, A.R., Martins, A.F., Paulino, A.T., Davi, M.F., Rubira, A.F., Muniz, E.C.: Superabsorbent hydrogels based on polysaccharides for application in agriculture as soil conditioner and nutrient carrier: a review. Eur. Polym. J. **72**, 365–385 (2015)
6. Sinha, S.: 14—Biodegradable superabsorbents: Methods of preparation and application—a review. In: Thomas, S., Balakrishnan, P., Sreekala, M.S. (eds.) Fundamental Biomaterials: Polymers, pp. 307–322. Woodhead Publishing, Cambridge, UK (2018)

7. Vasile, C., Pamfil, D., Stoleru, E., Baican, M.: New developments in medical applications of hybrid hydrogels containing natural polymers. Molecules **25**, 1539 (2020)
8. Chen, Y.: Hydrogels Based on Natural Polymers, p. 552. Elsevier, Amsterdam, The Netherlands (2020)
9. Mehr, M.J.Z., Omidian, H., Doroudiani, S., Kabiri, K.: Advances in non-hygienic applications of superabsorbent hydrogel materials. J. Mater. Sci. **45**, 5711–5735 (2010)
10. Jeddi, M.K., Laitinen, O., Liimatainen, H.: Magnetic superabsorbents based on nanocellulose aerobeads for selective removal of oils and organic solvents. Mater. Des. **183**(1), 08115 (2019)
11. Pérez-Álvarez, L., Ruiz-Rubio, L., Lizundia, E., Vilas-Vilela, J.L.: Polysaccharide-based superabsorbents: synthesis, properties, and applications. In: Mondal, M.I.H. (ed.) Cellulose-Based Superabsorbent Hydrogels, pp. 1393–1431. Springer Nature, Cham, Switzerland
12. Nayak, A.K., Hasnain, M.S., Pal, K., Banerjee, I., Pal, D.: Gum-Based Hydrogels in Drug Delivery. Biopolymer-Based Formulations, pp. 605–645. Elsevier, Amsterdam (2020)
13. Mignon, A., De Belie, N., Dubruel, P., Van Vlierberghe, S.: Superabsorbent polymers: a review on the characteristics and applications of synthetic, polysaccharide-based, semi-synthetic and 'smart' derivatives. Eur. Polym. J. **117**, 165–178 (2019)
14. Horie, K., Barón, M., Fox, R.B., He, J., Hess, M., Kahovec, J., Kitayama, T., Kubisa, P., Maréchal, E., Mörmann, W., et al.: Definitions of terms relating to reactions of polymers and to functional polymeric materials (IUPAC Recommendations 2003). Pure Appl. Chem. **76**, 889–906 (2004)
15. Bhatia, J.K., Kaith, B.S., Kalia, S.: Polysaccharide hydrogels: synthesis, characterization, and applications. In: Kalia, S., Sabaa, M.W. (eds.) Polysaccharide Based Graft Copolymers, pp. 271–290. Springer, Heidelberg, Germany (2013)
16. Boumalha, H.: Elaboration De Materiaux Composites Polymeres Superabsorbants/Additifs Et Etude Leurs Performances, Pour Une Application Dans Les Produits D'hygiene; University of Science and Technology Houari Boumediene: Bab Ezzouar, Algeria (2019)
17. Yabuki, A., Tanabe, S., Fathona, I.W.: Self-healing polymer coating with the microfibers of superabsorbent polymers provides corrosion inhibition in carbon steel. Surf. Coat. Technol. **341**, 71–77 (2018)
18. Qiao, D., Liu, H., Yu, L., Bao, X., Simon, G.P., Petinakis, E., Chen, L.: Preparation and characterization of slow-release fertilizer encapsulated by starch-based superabsorbent polymer. Carbohydr. Polym. **147**, 146–154 (2016)
19. Pourjavadi, A., Zeidabadi, F., Barzegar, S.: Alginate-based biodegradable superabsorbents as candidates for diclofenac sodium delivery systems. J. Appl. Polym. Sci. **118**, 2015–2023 (2010)
20. Ahmed, E.M.: Hydrogel: preparation, characterization, and applications: a review. J. Adv. Res. **6**, 105–121 (2015)
21. Rinaudo, M.: Main properties and current applications of some polysaccharides as biomaterials. Polym. Int. **57**, 397–430 (2008)
22. Pu, W., Shen, C., Wei, B., Yang, Y., Li, Y.: A comprehensive review of polysaccharide biopolymers for enhanced oil recovery (EOR) from flask to field. J Ind Eng Chem. **61**, 1–11 (2018)
23. Warson, H.: Modern superabsorbent polymer technology. Polym. Int. **49**, 1548 (2000)
24. Li, Q., Ma, Z., Yue, Q., Gao, B., Li, W., Xu, X.: Synthesis, characterization, and swelling behaviour of superabsorbent wheat straw graft copolymers. Bioresour. Technol. **118**, 204–209 (2012)
25. Fang, S., Wang, G., Li, P., Xing, R., Liu, S., Qin, Y., Yu, H., Chen, X., Li, K.: Synthesis of chitosan derivative graft acrylic acid superabsorbent polymers and its application as water retaining agent. Int. J. Biol. Macromol. **115**, 754–761 (2018)
26. Bierhalz, A.C.K., da Silva, M.A., Kieckbusch, T.G.: Natamycin release from alginate/pectin films for food packaging applications. J. Food Eng. **110**(1), 18–25 (2012)
27. Silva, N.H., Rodrigues, A.F., Almeida, I.F., Costa, P.C., Rosado, C., Neto, C.P., Freire, C.S., et al.: Bacterial cellulose membranes as transdermal delivery systems for diclofenac: in vitro dissolution and permeation studies. Carbohyd Polym. **106**, 264–269 (2014)

28. Hadrich, A.: Nouveaux Hydrogels A Base De Polysaccharide Obtenus Par Voie Biomimetique Ou Par Photoreticulation. University of Rouen Normandy, Rouen Normandy, France (2019)
29. Roberts, J.J., Martens, P.J.: 9—Engineering biosynthetic cell encapsulation systems. In: Poole-Warren, L., Martens, P., Green, R. (eds.) Biosynthetic Polymers for Medical Applications, pp. 205–239. Woodhead Publishing, Cambridge, UK (2016)
30. Zhang, J., Xiao, H., Li, N., Ping, Q., Zhang, Y.: Synthesis and characterization of superabsorbent hydrogels based on hemicellulose. J. Appl. Polym. Sci. **132**, 132 (2015)
31. Chang, C., Duan, B., Cai, J., Zhang, L.: Superabsorbent hydrogels based on cellulose for smart swelling and controllable delivery. Eur. Polym. J. **46**, 92–100 (2010)
32. Pourjavadi, A., Harzandi, A., Hosseinzadeh, H.: Modified carrageenan 3. Synthesis of a novel polysaccharide-based superabsorbent hydrogel via graft copolymerization of acrylic acid onto kappa-carrageenan in air. Eur. Polym. J. **40**, 1363–1370 (2004).
33. Shi, W., Dumont, M.-J., Ly, E.B.: Synthesis and properties of canola protein-based superabsorbent hydrogels. Eur. Polym. J. **54**, 172–180 (2014)
34. Liu, T., Wang, Y., Li, B., Deng, H., Huang, Z., Qian, L., Wang, X.: Urea free synthesis of chitin-based acrylate superabsorbent polymers under homogeneous conditions: E ects of the degree of deacetylation and the molecular weight. Carbohydr. Polym. **174**, 464–473 (2017)
35. Das, N.: Preparation methods and properties of hydrogel: a review. Int. J. Pharm. Pharm. Sci. **5**, 12–117 (2013)
36. Tolosa, L., Donato, M.T., Gómez-Lechón, M.J.: General cytotoxicity assessment by means of the MTT assay. In: Protocols in In Vitro Hepatocyte Research, pp. 333–348. Humana Press, New York, NY
37. Adler, S., Basketter, D., Creton, S., Pelkonen, O., Van Benthem, J., Zuang, V., Zaldivar, J.M.: Alternative (non-animal) methods for cosmetics testing: current status and future prospects—2010. Arch Toxicol. **85**(5), 367–485 (2011)
38. Loke, W.K., Lau, S.K., Yong, L.L., Khor, E., Sum, C.K.: Wound dressing with sustained anti-microbial capability. J. Biomed. Mater. Res. Off. J. Soc. Biomater. Jpn. Soc. Biomater. Aust. Soc. Biomater. Korean Soc. Biomater. **53**(1), 8–17 (2000)
39. Aday, A.N., Srubar, W.V.: 2—Biobased polymers for mitigating early- and late-age cracking in concrete. In: Pacheco-Torgal, F., Ivanov, V., Tsang, D.C.W. (eds.) Bio-Based Materials and Biotechnologies for Eco-Efficient Construction, pp. 19–41. Woodhead Publishing, Cambridge, UK (2020)
40. Jenkins, A.D., Kratochvíl, P., Stepto, R.F.T., Suter, U.W.: Glossary of basic terms in polymer science (IUPAC Recommendations 1996). Pure Appl. Chem. **68**, 2287–2311 (1996)
41. Kosemund, K., Schlatter, H., Ochsenhirt, J.L., Krause, E.L., Marsman, D.S., Erasala, G.N.: Safety evaluation of superabsorbent baby diapers. Regul. Toxicol. Pharmacol. **53**, 81–89 (2009)
42. Srinivas, S.M., Dhar, S.: Advances in diaper technology. Indian J. Paediatr. Dermatol. **17**, 83 (2016)
43. Munkvold, G.P.: Seed pathology progress in academia and industry. Annu Rev Phytopathol. **47**, 285–311 (2009)
44. El Hadrami, A., Adam, L.R., El Hadrami, I., Daayf, F.: Chitosan in plant protection. Mar. Drugs **8**(4), 968–987 (2010)
45. Biswas, M.C., Jony, B., Nandy, P.K., Chowdhury, R.A., Halder, S., Kumar, D., Imam, M.A., et al., Recent advancement of biopolymers and their potential biomedical applications. J. Polym. Environ. 1–24 (2021)
46. Sinha, S.: Textile wastewater clarification using milk based formulations as an effective dye absorbent and flocculant. Mater. Today Proc., Elsevier **67**, 1304–1309 (2022)
47. Sinha, S.: Biopolymer derived superabsorbent for environmental sustainability: a review. Environ. Q. Manage., Wiley **32**, 177–185 (2022)
48. Sinha, S.: Derivatization of milk protein using poly (acrylamide): its characterization and application. Macromol. Symp. **381**(1), 1800120 (2018)
49. Queen, D., Gaylor, J.D.S., Evans, J.H., Courtney, J.M., Reid, W.H.: The preclinical evaluation of the water vapour transmission rate through burn wound dressings. Biomaterials **8**(5), 367–371 (1987)

Bio-based Versus Petro-based Superabsorbent Polymers

Shiv Kumari Panda

Abstract Superabsorbent polymers (SAPs) are simply three-dimensional network types of hydrophilic materials which have the propensity to soak up and conserve huge capacities of water or aqueous solutions. They can intake very high quantity of water as hundreds to thousands times more extent in comparison to its original weight within a certain time period without suffering any break up. Generally, SAPs are white coloured sugar like hygroscopic substance. Broadly, they have enormous applications in disposable diapers, female personal care products and also in the agricultural field. Due to the presence of different types of monomers and varying polymeric units, SAPS are categorized as various types. Broadly, they are man-made or petrochemical-based and natural polysaccharide or polypeptide-based. Main constituent of current synthetic superabsorbent are based on acryl amide, acrylic acid or its salts with various types of initiator, cross-linker and promoter through the solution suspension polymerization technique. But various problems associated with such synthetic superabsorbent are poor biodegradability and high costs. So to overcome these bottlenecks, there is a growing interest towards the bio-based superabsorbent for getting the solution for various environmental issues.

Keywords Superabsorbent polymers (SAPs) · Petro-based · Bio-based · Polysaccharide · Poly acrylate

1 Introduction

The gist of twelve principles of green chemistry was based on the concept that prevention of waste or perfect utilization of waste as another raw material for a new synthesis is more acceptable than cleaning up the same. So according to this motto, the current situation requires to differentiate among bio-based material and synthetic material regarding their behaviour towards our environment. Actually in

S. K. Panda (✉)
Department of Chemistry, U.N. Autonomous College of Science & Technology, Adaspur, Cuttack, Odisha, India
e-mail: shivkumaripanda@gmail.com

S. Pradhan and S. Mohanty (eds.), *Bio-based Superabsorbents*, Engineering Materials,
https://doi.org/10.1007/978-981-99-3094-4_3

our day-to-day life, we are used to dealing with various types of materials that are made up of plastic. These materials are widely used by small children as well as the young generation. So starting from toys to water bottles or various fast consumer-moving goods to biomedical products, all materials have special importance in their unique way for different age groups. Before going to discuss their utilization, we the responsible educated mass have an important duty to discuss the starting material from which they have generated. Are they renewable and bio-based or non-renewable and petroleum-based? Today, the term gel is used by all age group people as it has potential applications in the toy and cosmetic industry. In hydrophilic condition, these gels can be referred as hydrogel. This hydrogel has become a hot research issue as it is the mother of all types of superabsorbent polymers (SAPs). Generally, SAPs are three-dimensional, covalently bonded, cross-linked networks of synthetic or natural monomer-based swollen structures [1]. They are white coloured sugar like material. Basically, they have amorphous or crystalline or semi-crystalline structure with a maximum capacity of water absorption characteristics [2, 3]. The ideal nature of SAPs is based on the characteristics like high absorption tendency, high gel fraction after cross-linking, inexpensiveness, excellent durability and stability upon swelling or during storage, non-toxicity and re-wetting capability [4–7]. Figure 1 shows the general mechanism of superabsorbent. These SAPs has been broadly used in biomedical [8], cosmeceutical [9], agricultural [10, 11] and personal healthcare applications. But more than 90% of SAPs are synthesized from non-renewable petroleum-based chemical material. In very ancient times, the most used water-absorbing substances are fibres like tissue paper, cotton, sponge, etc. Such materials have a water storing capacity of approximately 15–20 times their original weight. But, for the first time superabsorbent polymers (SAPs) were discovered in the laboratory of the United States Agricultural Department in the year 1960, and the aim was to enhance the water preservation capacity of the soil. These are similar to sodium poly acrylate and are named as "super slurper". Then a revolution started in this industry and a variety of technologies have been initiated by taking the grafting technique with various chemicals like acrylic acid, acryl amide, polyvinyl alcohol, etc. In the early seventies, SAPs came to market for the first time in favour of commercial purposes. At that time also, the purpose was not only limited to agricultural use but also to other important applications like personal healthcare products including disposable hygienic materials. In the year 1978, Park Davis used superabsorbent polymers in sanitary napkins.

Typically, petro-based or synthetic SAPs can be developed by using the chemicals like acrylic acid, dimethylaminoethyl methacrylate, silicone, acryl amide, 2-acrylamido-2-methylpropane sulfonic acid, poly ethylene glycol, dimethyl amino-propyl methacrylamide, methacrylic acid, etc. The advantages of these types of SAPs are their purity, abundance and maximum applicability. But they are non-renewable, toxic and lack control over their rate of reaction. So to overcome such types of bottlenecks, the introduction of natural SAPs became necessary for the current situation. They include starch, alginate, cellulose, lignin, gelatin, chitosan, pectin, guar gum, etc. By nature, these are eco-friendly, renewable, easily available and highly

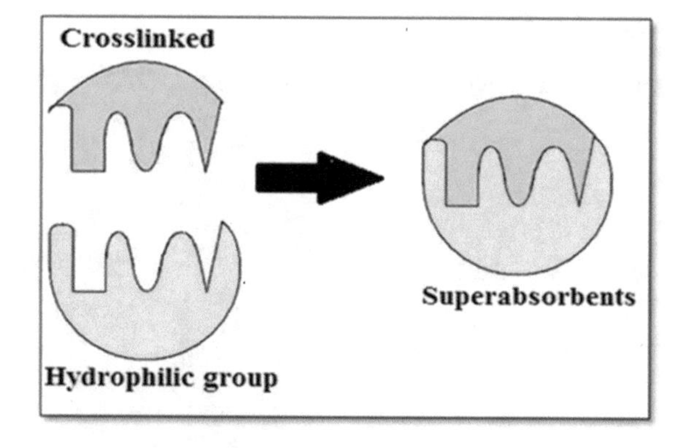

Fig. 1 From cross-linking of hydrophilic monomer to superabsorbent

adhesive due to their inherent tendency. But for these natural SAPs, various extraction methods are followed for finding the starting reactant material and also they need some modification to enhance their storage shelf-life, as they are very sensitive in nature. Presently, various semi-synthetic SAP materials are also available with superior characteristics to both synthetic and natural polymers. Examples of such hybrid types of SAPs are starch-poly acrylamide, chitin acrylate, acrylic acid gelatin, carboxymethyl cellulose-acrylic acid, hydroxyethylacryl chitosan-sodium alginate and cellulose-polyethylenimine. But they possess negative properties like phase separation, loss of biodegradability and homogeneity.

2 Difference Between Absorbing and Super-Absorbing Material

All the absorbing and super-absorbing materials are hygroscopic in nature. These hygroscopic substances are generally differentiated into two main groups based on their pattern of absorption of liquid or solvent. This mechanism of absorption may be physical or chemical type. Chemical absorption proceeds through chemical reaction by completely altering the whole identity of the starting material. Physical absorption occurs through the mechanism like reversible changes of the crystal structure or physical entrapment of water through capillary force [12]. Actually, traditional absorbent materials absorb liquid solvent or water similar to superabsorbent polymers. But they release the absorbed water when they are squeezed. Such materials are like tissue papers or cotton. On the other hand, superabsorbent polymers are an organic type of material with the extraordinary ability of water absorption. They can absorb like hydrogels, but retain the solvent inside them for a long period of time. Figure 2 represents the schematic illustrations of a superabsorbent polymer-based hydrogel in both dry and water-swollen states. This figure demonstrated that the cross-linked chains are near to each other and the pores are filled by water. Again the functional groups are in close proximity to each other because of hydrogen bonding.

Fig. 2 Schematic illustrations of a superabsorbent polymer

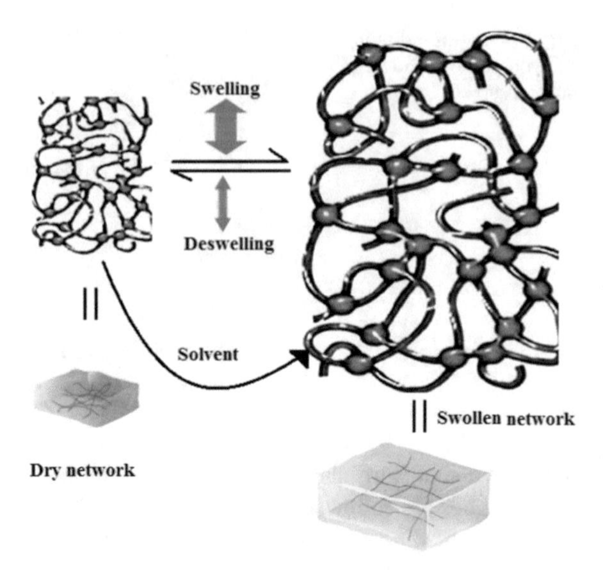

When water enters the system, then the pores as well as the polymeric chains get fully filled and expanded. The whole structure gets swelling and converts to a bigger size as all the pores are filled with water.

3 Classification of Superabsorbent Polymer

Scheme 1 gives a solid idea regarding the classification of SAPs from various points of view. Morphologically, SAPs represent different types of categorization with various challenging applications. These structures include fibres [13, 14], powders [15, 16], granules [17, 18] and sheets [19]. Actually, the original shapes of SAPs are always maintained even after water uptake. This implies that the SAPs have enough strength to rule out any physical degradation even upon exposure to pressure with no effect on structure [20]. Based on the cross-linking mechanism, this classification may be physical and chemical type. Physical cross-linking is reversible in nature and chemical cross-linking is permanent in nature. This category is based on association mechanisms of the participating polymer chains.

Moreover, physical bonds are based on weak forces of attraction like hydrogen bonding, whereas chemical bonds represent strong forces of attraction like covalent bonding and proceed through free radical polymerization mechanism. Also, physically cross-linked SAPs have limitations, as very particular class of polymeric materials could be cross-linked by this process. Chemical cross-linking is the most acceptable one for major types of polymeric materials in the presence of initiators

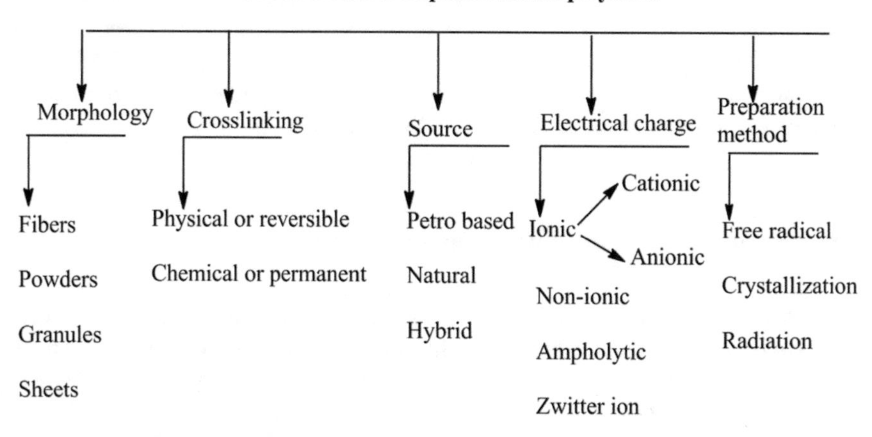

Scheme 1 Classification of superabsorbent polymers

and cross-linking agents. Furthermore, physically cross-linked SAPs are inhomogeneous networks as the chain entanglements form rough and uneven porous structures, while chemically cross-linked SAPs are covalently cross-linked arrangements forming regular porous structures [21]. Classification based on building block or source showed that the SAPs are either being synthetic or natural or hybrid/semi-synthetic type [22, 23]. Petro-based or synthetic SAPs are generally synthesized from petrochemical-based monomers such as acrylates or acryl amides [24–26], while natural SAPs include polypeptides and polysaccharides [27, 28]. In case of hybrid SAPs, both natural and synthetic polymer sources are generally combined to create SAPs that exhibit superior characteristics towards specialized applications [29]. This classification is the most general and broadly used one. Now classification based on the type of electrical charges represents four categories depending on the presence of electrical charges along the polymer backbone and side chains. These may be non-ionic polymeric types possessing no charges, ionic SAPs with either anionic or cationic groups, ampholytic with both acidic as well as basic functionalities or zwitter ionic SAPs containing both anionic and cationic groups with an overall net charge of zero. This classification is used in the perspective of electro-sensitive "smart" SAPs [29].

4 Types of Superabsorbent Polymers Based on Starting Material

Relating to previous sections, the most considerable subdivisions of SAPs are based on their types of starting material from which they are originated. In the context of the building block or source of SAPs, it is of different types like petro-based, natural and semi natural or hybrid type of superabsorbent polymers [30]. The petro-based SAPs are mechanically robust and stiffer compared to natural polymers. But mechanical strength of natural polymers is highly poor. Hybrid types of SAPs have advanced properties as these are the combination of both synthetic and natural type. The detailed discussions of all types of SAPs are as follows.

4.1 Petro-based Superabsorbent Polymers

In the field of superabsorbent polymers, majority of SAPs used these days have a synthetic or petrochemical-based origin. In industries, they are basically synthesized from the monomeric units of acrylic acid, sodium or potassium salts of acrylic acid and acrylamide. One problem associated with this acrylic acid monomer is that during its storage, inhibitor methoxy hydroquinone is used to avoid the tendency of spontaneous self-polymerization. But in industries, these inhibitors are not separated during the polymerization process due to some technical and cost-related issues associated with these processes [31], as a result of which unnecessary by-products are released in dimer form in the immensity of acrylic acid through Michael addition reaction. So the production of this dimeric acrylic acid should be checked to stop the adverse effect of the resulting product as it possess some environmental issues and is hazardous for the ecosystem [32]. Other types of monomers used in various laboratories for the development of petro-based SAPs are methacrylic acid, methacryl amide, acrylo nitrile, 2-hydroxyethyl methacrylate, 2-acrylamido-2-methylpropane sulphonic acid, N-vinyl pyrrolidone, vinyl sulphonic acid, vinyl acetate, etc. [20]. But most used monomers are acrylic acid and acrylamide [33]. Structure of some types of synthetic SAPs are given in Scheme 2.

All the monomers are introduced into water-soluble cross-linkers like N, N'-methylene bisacrylamide followed by the solution polymerization technique. But such reactions involve some negative issues like lack of control over the kinetics of reaction and particle size distribution that leads to unwanted solid or rubbery products. Petro-based SAPs have high applicability in the field of biomedical, and it can be used as coatings for catheters [34, 35], drug delivery processes [8, 36], dressings of burn [37, 38], preparations of various electrophoresis gels [39] and in the ground of agriculture and forestry also [40]. They have also non-biomedical applications in gardening and manufacturing of nursing mats, medical bandages, diapers and female hygiene products. It can be used as water beads for plants [41] in various types of water purification systems [42]. Modern applications of SAPs are based on synthetic polymers like acrylic acid and acryl amide in concrete to alleviate the auto shrinkage [43] or self-healing applications [19]. Again acryl amide-related SAPs have proved their potential utilization in agricultural fields [44]. Poly (acrylate/acrylic acid)-based SAPs can be used for water conservation in sandy soil [44]. Even though synthetic SAPs have proved their candidature in various fields, one of the demerits is its non-biodegradability. So, the more acceptable renewable approach is

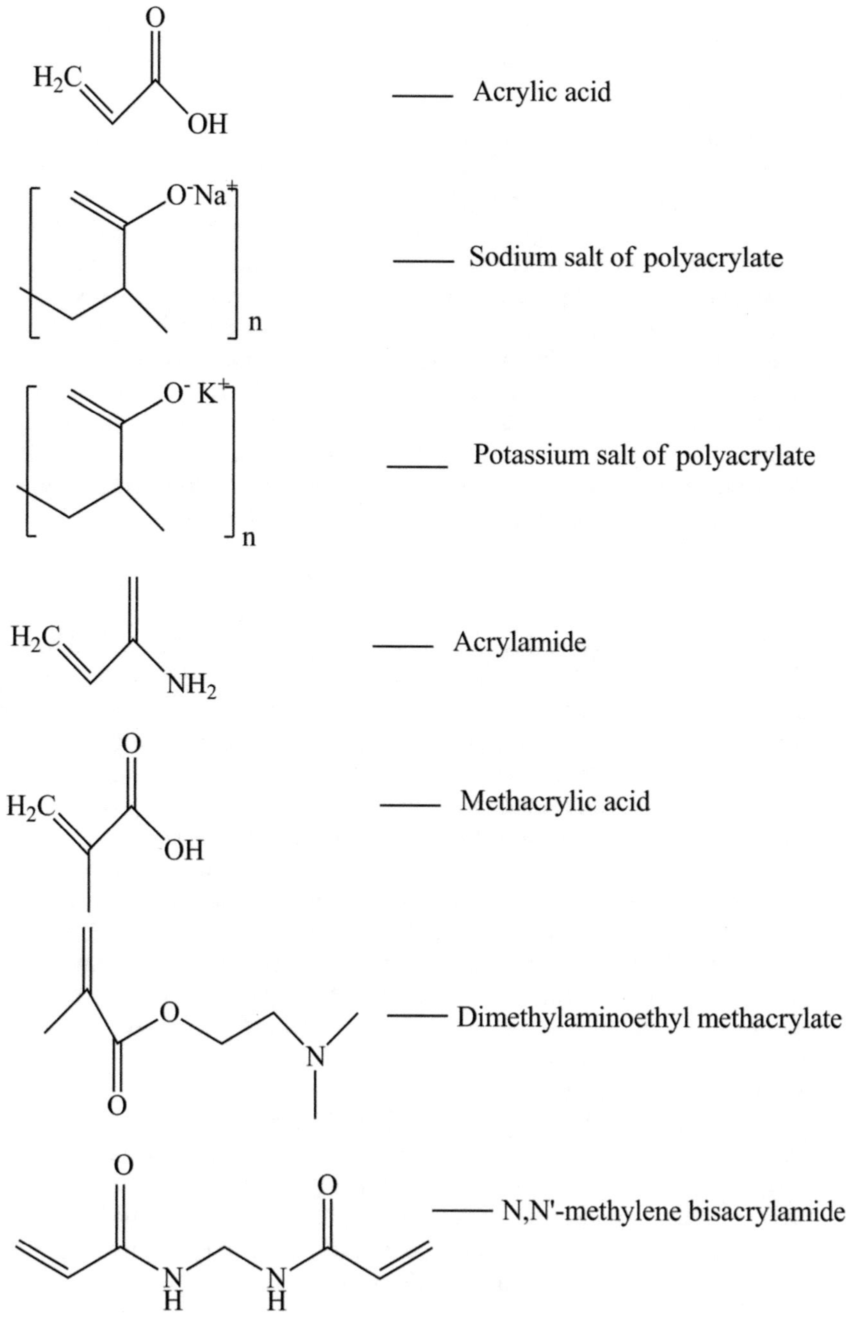

— Acrylic acid

— Sodium salt of polyacrylate

— Potassium salt of polyacrylate

— Acrylamide

— Methacrylic acid

— Dimethylaminoethyl methacrylate

— N,N'-methylene bisacrylamide

Scheme 2 Structure of synthetic types of superabsorbent polymers based on the use of natural SAPs or hybrid SAPs. Some major types of SAPs are discussed in the following section.

(a) **Acrylic acid**
Acrylic acid is a simple colourless liquid of carboxylic acid derivative. It is a weak organic acid and its IUPAC name is 2-propenoic acid. This is mainly utilized for the synthesis of acrylic-based esters as well as resins which are main ingredients in coatings and adhesives industry. For the polymerization of acrylic acid monomer, N, N'-methylene bisacrylamide is used as an efficient cross-linker followed by the solution polymerization technique. It has successfully proved its applicability as detergent intermediates and in oil and water treatment too. The most important application of this acrylate monomer is that, in the polymeric form, i.e. in the form of poly acrylic acid, it can behave as a superabsorbent polymer. Acrylic acid is used in the manufacturing of plastics and paint. But the Environmental Protection Agency of United States has declared acrylates as human carcinogenic. They have strong irritating effect towards human eye, skin and mucous membrane on direct contact.

(b) **Sodium poly acrylates**
It is the sodium salt of poly acrylic acid and has major applications in consumer products. It is also famously known as "water lock" due to its superabsorbent tendency. In the early sixties, Department of Agriculture of the United States developed the first SAPs materials similar to this sodium poly acrylates. Generally, SAPs used in disposable hygiene products; diapers and sanitary napkins are typically based on sodium poly acrylate. Due to its superabsorbent tendency, it can absorb the solvent thousand times more than the original weight. Sodium poly acrylate is an anionic polyelectrolyte having negatively charged carboxylic functional moiety in the chain backbone. Sodium poly acrylate is a chemical polymer that contains chains of acrylate compounds. Here, the presence of sodium positively enhances the ability of absorption of large amounts of water. When dissolved in water, it forms a thick and transparent solution due to the ionic interactions of the molecules [45]. Sodium poly acrylate has many attractive mechanical properties and good thermal stability with strong hydration capacity [45]. It can be used as a potential foodstuff preservative for bread, juice and ice cream also.

(c) **Potassium poly acrylate**
It is the potassium salt of poly acrylic acid and can be used as water-preserving mediator in agricultural sector and improves plant growth by diminishing the obstruction factors. It also soaks up and liberates soil nutrients. It can act as water-soluble chemical and fertilizer in a similar way as water, by building a well microenvironment in the plant root area. So these SAPs of potassium poly acrylate have greater applicability in the agricultural sector.

(d) **Acrylic amide**
Acrylic amide is a white coloured powder like hydrophilic, cheap, vinyl monomer-based pure organic compound that can be used without purification.

It is a volatile type of hazardous vinyl substituted primary amide. As a superabsorbent polymeric material, it can form a gel-like structure after absorbing any particular solvent. It is soluble in water as well as in several organic solvents. Polymerization of acryl amide leads to the formation of hydrophilic chains by improving the water absorption tendencies of the whole system [46]. Majority of acrylic amide is industrially synthesized in the form of precursor to poly acryl amides, which has broad applicability as water-soluble thickeners and flocculation agents [47]. Actually, acrylic amide is highly toxic whereas its polymeric form, i.e. poly acrylmide is non-hazardous in nature. But continuous use of such products is hazardous for the environment and human beings.

In spite of the broad utilization of the aforementioned synthetic absorbent polymers in a variety of fields, they are associated with severe environmental issues like poor biodegradability as well as high cost [48]. As a result of this, there is a rising interest to build up natural SAPs that have the capacity to conquer these drawbacks, showing both immense water uptake capability with easy processability.

4.2 *Bio-based Superabsorbent Polymers*

The family of a majority of eco-friendly and renewable SAPs constitutes polysaccharides, starch and proteins. They basically originated from plants and animals. Broadly used Polysaccharides are the potential biosynthetic derivatives of plants and animals. Bio-based SAPs have inherent properties like non-toxicity, renewability, easy availability, inexpensiveness, bio-friendliness and sustainability. Undeniably, such green, benevolent, safer bio-based SAPs have no environmental dilemma in comparison to synthetic SAPs. Again, they have no contribution towards the formation of any waste or by-product after their utilization. According to current report, polysaccharides are also derived from bacteria like bacterial hyaluronan and gellan gum [49]. Scheme 3 gives an idea regarding the types of bio-based SAPs. Mostly used family of natural polymers for SAPs cover the polysaccharides that are may be fish and collagen -based or like alginate, chitosan, agar, carrageenan, dextrin, cellulose, starch, xanthan and proteins [30, 32, 50–56]. Water-soluble polysaccharides contain functional moieties like alcohols, carboxylic acids, amine, etc. Basically, such types of substituents can easily participate in cross-linking or grafting with other polymers. Also, they can easily form intermolecular hydrogen bonding with other groups present near them. So based on these reasons, it can be strongly assumed that natural polymers have greater demand as SAPs in comparison to artificial synthetic one. So, such bio-based SAPs are gaining more and more attention than the contemporary petro-based non-renewable ones.

Actually, majority portion of the market of superabsorbent materials are based on petrochemical derivatives. In petro-based superabsorbent polymers, acrylic acid plays a very important role due to its affordable price to efficiency balance [57]. But from the environmental point of view, it is necessary to give first priority to "greener"

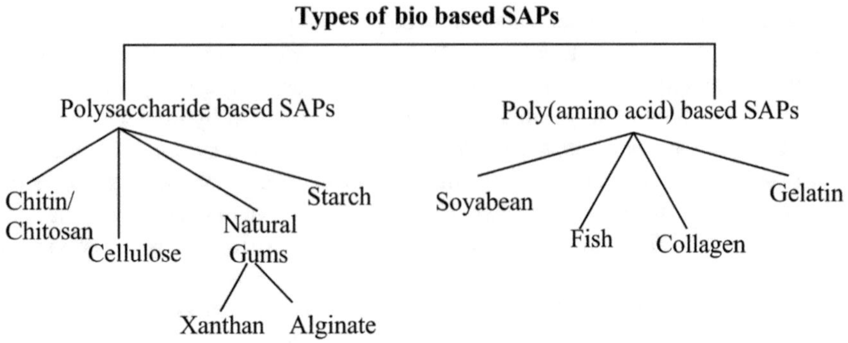

Types of bio based SAPs

- Polysaccharide based SAPs
  - Chitin/Chitosan
  - Cellulose
  - Natural Gums
    - Xanthan
    - Alginate
  - Starch
- Poly(amino acid) based SAPs
  - Soyabean
  - Fish
  - Collagen
  - Gelatin

Scheme 3 Major types of bio-based superabsorbent polymers

products [58]. Such type of mindset can drive the attention of current scenario towards bio-based superabsorbents. According to the current situation, biodegradable waste products can be chosen for the preparation of SAPs. Following sections of the article can give some idea regarding the types of natural polymers available for the synthesis of SAPs.

(a) **Polysaccharide-based bio SAPs**

Polysaccharide-based superabsorbent polymers are the inexpensive, easily available and biodegradable organic materials. Some of the well-known examples of polysaccharide SAPs are chitin, cellulose, starch and natural pH-sensitive gum like xanthan and alginates. Usually, two different types of reactions are followed for the synthesis of polysaccharide-based SAPs. These may be graft copolymerization of polysaccharide monomer in the presence of cross-linker, or direct cross-linking of polysaccharide units.

Chitosan is a linear polysaccharide derived from starting material chitin through the deacetylation process and the main building block is glucosamine. It has very good thermal stability. It is a green polymer and it has poly-cationic characteristics with multiple interesting inherent antibacterial and antifungal qualities. It is biocompatible with a number of organic and inorganic compounds due to the availability of free $-NH_2$ and –OH functional moieties in it. Again, $-NH_2$ functional moieties are more reactive than –OH groups [59]. These groups can easily attach with other functional groups by certain electrostatic forces of attraction or by intermolecular hydrogen bonding. Such cationic nature of chitosan makes it capable to attach to the negatively charged end of bacterial cell membranes. Again, chitosan is soluble in dilute acid solutions and can be fixed with polyanions to form complexes and gels. So specifically, the presence of free $-NH_2$ and –OH groups on chitosan can prove it as a potential candidate to transfer it to various interesting forms in a unique way. It has also broad applicability as superabsorbent in the polymer industry. Nowadays, for increasing the applicability of chitosan in the SAP industry, various modification like carboxy methylation, phosphorylation and alkylation of chitosan have been performed at the N– and O– positions of the $-NH_2$ and –OH groups [60, 61]. N– and O– carboxymethyl chitosan has potential application in haemostatic wound dressings [62].

Chitosan on reaction by oxalic acid produces a superabsorbent-based hydrogel that has the capacity to undergo physical cross-linking through an ammonium carboxylate complex that can effectively absorb copper(II) [62].

Alginate is one of the most used polar, unbranched, anionic polysaccharide collected from the cell walls of brown algae [63, 64]. The quality and nature of alginate depends on the type and age of the algae and the applied technical methods [65–67]. It is also commercially available as a sodium salt of alginate (NaAlg) [68, 69]. The accurate composition of alginate chains differ according to the source from which they are collected, the harvest location, the season and the fraction of the seaweed used. These parameters also affect the gelling tendency and potency of the formed alginate. Alginate contains carboxylic acid groups which become negatively charged in aqueous solutions. It can be manufactured by two types of processing techniques such as "acid precipitation method" and the "calcium precipitation method". Interestingly, when sodium alginate is combined with salts like calcium chloride then the divalent Ca^{2+} ions from the salt form a helical chain-like structure, and the whole system was converted to an "egg box" like model [69]. Typically, this is a cross-linked network as the carboxylate functionalities make coordinate bond with the cations, and the entire structure turn water-insoluble. On the other hand, the anionic moieties will attract water into the structure and showed SAP-like behaviour. Alginate has already proved its candidature for biomedical applications particularly for controlled drug release, cell encapsulation, dental impression, wound dressing [70] and in bio plastics for packaging, textiles and paper industries. It is also used in the food industry as a stabilizer, emulsifier and gelling agent.

Cellulose is the most available, linear homo polysaccharide-based organic compound that is composed of β-1, 4-anhydro-D-glucose moieties and derived from biomass [71]. The main sources of cellulose are wood and plant fibres, leaf, stalk, fruit, etc. Here, the presence of the most active hydroxyl groups can make strong intermolecular hydrogen bonding. It has also unique cohesive nature. Due to such tendencies, this cellulose can be used as SAPs by various industries. Celluloses have already been successfully introduced in the field of paper, textile and the material industry. It has also greater applicability in the industry of diaper and various hygiene product materials. Modification of cellulose to carboxymethyl cellulose has better application as SAPs in biomedical field.

Starch is the second most abundant natural biomass produced from plant roots, stalks and crop seeds. The major sources of starch are potatoes, wheat and maize [72]. It is the earliest commercial superabsorbent polymer. It is composed of glucose units connected by glycosidic bonds with linear amylose and branched amylopectin functional moieties. These amylopectin groups enhance the viscosity of the polymeric chain backbone, and amylose segments enhance the grafting and swelling efficiency [73]. It has industrial applications for the production of alcohols and bio fuels. It can also be useful as a thickening or glueing agent [74].

5 Conclusion

Actually, the beautiful world has already been ruined by various unenthusiastic activities of mankind. Although both synthetic and bio-based superabsorbent polymers have already been used for various applications like diapers, the biomedical field, agriculture, etc. with respect to cost-effectiveness, renewability, non-toxicity, bio degradability and sustainability, the natural SAPs are of unique interest. So now it is high time for the natural re-establishment of these lost assets, which is standing as a major challenge for human society. The journey towards sustainability can be better accompanied and progressed by focusing on the utilization of bio-based materials rather than artificial synthetic materials. Again, blending of synthetic monomers with polymers of bio-based origin will also improve the sustainability of synthetic SAPs. As synthetic SAPs are still used in a major part of current applications, "smart" semi-synthetic SAPs require additional research efforts in the upcoming years.

References

1. Ratner, B.D., Hoffman A.S.: Hydrogels for Medical and Related Applications. Andrade, JD, Washington, DC, American Chemical Society (1976)
2. Park, K., Shalaby, W.S.W., Park, H.: Biodegradable Hydrogels for Drug Delivery. Technomic Publishing Company, Inc., Basel (1993)
3. Anderson, J.M.: Polymeric Biomaterials. NATO ASI Series. Springer (1986)
4. Yang, S.T.; Park, Y.S.: Release pattern of dexamethasone after administration through an implant-mediated drug delivery device with an active plunger of super absorbent polymer. Drug Deliv. Transl. Res. (2018)
5. Fahmy, T.Y.A., Mobarak, F.: Green nanotechnology: a short cut to beneficiation of natural fibers. Int. J. Biol. Macromol. (2011)
6. Yang, L., Yang, Y., Chen, Z., Guo, C., Li., S.: Influence of super absorbent polymer on soil water retention, seed germination and plant survivals for rocky slopes eco-engineering. Ecol. Eng. (2014)
7. Alharbi, K., Ghoneim, A., Ebid, A., El-Hamshary, H.; El-Newehy, M.H.: Controlled release of phosphorous fertilizer bound to carboxymethyl starch-g-polyacrylamide and maintaining a hydration level for the plant. Int. J. Biol. Macromol. (2018)
8. Ahmed, E.M.: Hydrogel: preparation, characterization, and applications: a review. J. Adv. Res. (2015)
9. Bukhari, S.M.H., Khan, S., Rehanullah, M., Ranjha, N.M.: Synthesis and characterization of chemically cross-linked acrylic acid/gelatin hydrogels: effect of pH and composition on swelling and drug release. Int. J. Polym. Sci. (2015)
10. Mignon, A.: Effect of pH-responsive superabsorbent polymers on the self-sealing and self-healing of cracks in concrete. Ghent University, PhD (2016)
11. Mohammad, K.K., Zohuriaan Mehr, M.J.: Superabsorbent polymer materials: a review. Iran. Polym. J. (2008)
12. Zohuriaan Mehr, M.J.: Super-absorbents (in Persian). Iran Polym. Soc. (2006)
13. Yabuki, A., Tanabe, S., Fathona, I.W.: Self-healing polymer coating with the microfibers of superabsorbent polymers provides corrosion inhibition in carbon steel. Surf. Coat. Technol. (2018)
14. Islam, M.S, Rahaman, M.S., Yeum, J.H.: Electrospun novel super-absorbent based on polysaccharide–polyvinyl alcohol–montmorillonite clay nanocomposites. Carbohydr. Polym. (2015)

15. Mignon, A., Graulus, G.J., Snoeck, D., Martins, J., De Belie, N., Dubruel, P., Van Vlierberghe, S.: pH-sensitive superabsorbent polymers: a potential candidate material for self-healing concrete. J. Mater. Sci. (2014)
16. Pelto, J., Leivo, M., Gruyaert, E. Debbaut, B., Snoeck, D., De Belie, N.: Application of encapsulated superabsorbent polymers in cementitious materials for stimulated autogenous healing. Smart Mater. Struct. (2017)
17. Mudiyanselage, T.K., Neckers, D.C.: Highly absorbing superabsorbent polymer. J. Polym. Sci. Part A: Polym. Chem. (2008)
18. Qiao, D., Liu, H., Yu, L., Bao, X., Simon, G.P., Petinakis, E., Chen, L.: Preparation and characterization of slow-release fertilizer encapsulated by starch-based superabsorbent polymer. Carbohydr. Polym. (2016)
19. Mignon, A., Snoeck, D., Schaubroeck, D., Luickx, N., Dubruel, P., Van Vlierberghe, S., De Belie, N.: pH-responsive superabsorbent polymers: a pathway to self-healing of mortar. React. Funct. Polym. (2015)
20. Nesrinne, S., Djamel, A.: Synthesis, characterization and rheological behavior of pH sensitive poly (acrylamide-co-acrylic acid) hydrogels. Arab. J. Chem. (2017)
21. Hoffman, A.S.: Hydrogels for biomedical applications. Adv. Drug Deliv. Rev. (2012)
22. Van den Heede, P., Mignon, A., Habert, G., De Belie, N.: Cradle-to-gate life cycle assessment of self-healing engineered cementitious composite with in-house developed (semi-) synthetic superabsorbent polymers. Cem. Concr. Compos. (2018)
23. Chaithra, G.M., Sridhara, S.: Growth and yield of rainfed maize as influenced by application of super absorbent polymer and Pongamia leaf mulching. IJCS (2018)
24. Zhang, B., Cui, Y., Yin, G., Li, X., Liao, L., Cai, X.: Synthesis and swelling properties of protein-poly (acrylic acid-co-acrylamide) superabsorbent composite. Polym. Compos. (2011)
25. Liu, T., Wang, Y., Li, B., Deng, H., Huang, Qian Z.L., Wang, X.: Urea free synthesis of chitin-based acrylate superabsorbent polymers under homogeneous conditions: effects of the degree of deacetylation and the molecular weight. Carbohydr. Polym. (2017)
26. Fang, S., Wang, G., Li, P., Xing, R., Liu, S., Qin, Y., Yu, H., Li, X.K.: Synthesis of chitosan derivative graft acrylic acid superabsorbent polymers and its application as water retaining agent. Int. J. Biol. Macromol. (2018)
27. Qiao, D., Yu, L., Bao, X., Zhang, B., Jiang, F.: Understanding the microstructure and absorption rate of starch-based superabsorbent polymers prepared under high starch concentration. Carbohydr. Polym. (2017)
28. Graulus, G.J., Mignon, A., Van Vlierberghe, S., Declercq, H., Fehér, K., Cornelissen, M., Martins, J.C., Dubruel, P.: Cross-linkable alginate-graft-gelatin copolymers for tissue engineering applications. Eur. Polym. (2015)
29. Pande, P.: Polymer hydrogels and their applications. Int. J. Mater. Sci. (2017)
30. Mohammad, K.K., Zohuriaan-Mehr, J.: Superabsorbent polymer materials: a review. Iran. Polym. J. (2008)
31. Pourjavadi, A., Kurdtabar, M., Mahdavinia, R.G., Hosseinzadeh, H.: Synthesis and superswelling behavior of a novel protein-based superabsorbent hydrogel. Polym. Bull. (2006)
32. Buchholz, F.L., Graham, A.T.: Modern superabsorbent polymer technology. Wiley-VCH, New York (1998)
33. El-Tohamy, W.A., El-Abagy, H.M., Ahmed, E.M., Aggor, F.S., Hawash, S.I.: Application of super absorbent hydrogel poly (acrylate/acrylic acid) for water conservation in sandy soil. Transac. Egypt. Soc. Chem. Eng. (2014)
34. Hanak, B.W., Hsieh, C.Y., Donaldson, W., Browd, S.R., Lau, K.K., Shain, W.: Reduced cell attachment to poly (2-hydroxyethyl methacrylate)-coated ventricular catheters in vitro. J. Biomed. Mater. Res. B Appl. Biomater. (2018)
35. Ahearn, G., Grace, D.T., Jennings, M.J., Borazjani, R.N., Boles, K.J., Rose, L.J., Simmons, R.B., Ahanotu, E.N.: Effects of hydrogel/silver coatings on in vitro adhesion to catheters of bacteria associated with urinary tract infections. Curr. Microbiol. (2000)
36. Hoare, T.R., Kohane, D.S.: Hydrogels in drug delivery: progress and challenges. Polymer (2008)

37. Mogoşanu, G.D., Grumezescu, A.M.: Natural and synthetic polymers for wounds and burns dressing. Int. J. Pharm. (2014)
38. Kamoun, E.A., Kenawy, E.R.S., Chen, X.: A review on polymeric hydrogel membranes for wound dressing applications: PVA-based hydrogel dressings. J. Adv. Res (2017)
39. Kubo, T., Nishimura, N., Furuta, H., Kubota, K., Naito, T., Otsuka, K.: Tunable separations based on a molecular size effect for biomolecules by poly (ethylene glycol) gel-based capillary electrophoresis. J. Chromatogr. A (2017)
40. Huettermann, A., Orikiriza, L.J., Agaba, H.: Application of superabsorbent polymers for improving the ecological chemistry of degraded or polluted lands. CLEAN—Soil, Air, Water (2009)
41. Umachitra, G.: Disposable baby diaper–a threat to the health and environment. J. Environ. Sci. Eng. (2012)
42. La, Y.H., McCloskey, B.D., Sooriyakumaran, R., Vora, A., Freeman, B., Nassar, M., Hedrick, J., Nelson, A., Allen, R.: Bifunctional hydrogel coatings for water purification membranes: improved fouling resistance and antimicrobial activity. J. Membr. Sci. (2011)
43. Schrofl, C., Mechtcherine, V., Gorges, M.: Relation between the molecular structure and the efficiency of superabsorbent polymers (SAP) as concrete admixture to mitigate autogenous shrinkage. Cem. Concr. Res. (2012)
44. El Tohamy, W.A., El-Abagy, H.M., Ahmed, E.M., Aggor, F.S., Hawash, S.I.: Application of super absorbent hydrogel poly (acrylate/acrylic acid) for water conservation in sandy soil. Trans. Egypt. Soc. Chem. Eng. (2014)
45. Yang, Y., Rengui, P., Cheng, Y., Youhong, T.: Eco-friendly and cost-effective superabsorbent sodium polyacrylate composites for environmental remediation. J. Mater. Sci. (2015)
46. Qin, Q., Tang, Q., Li, Q., He, B., Chen, H., Wang, X., Yang, P.: Incorporation of H_3PO_4 into three-dimensional polyacrylamide-graft-starch hydrogel frameworks for robust high-temperature proton exchange membrane fuel cells. Int. J. Hydrogen Energy (2014)
47. Shi, W., Dumont, M.J., Ly, E.B.: Synthesis and properties of canola protein-based superabsorbent hydrogels. Eur. Polym. (2014)
48. Kim, Y.J., Yoon, K.J., Ko, S.W.: Preparation and properties of alginate superabsorbent filament fibers crosslinked with glutaraldehyde. J. Appl. Polym. Sci. (2000)
49. Dutkiewicz, J.K.: Superabsorbent materials from shellfish waste—a review. J. Biomed. Mater. Res. (2002)
50. Wang, J., Mignon, A., Trenson, G., Van Vlierberghe, S., Boon, N., De Belie, N.: A chitosan based pH-responsive hydrogel for encapsulation of bacteria for self sealing concrete. Cem. Concr. Compos. **93**, 309–322 (2018)
51. Panda, S.K.: Synthesis and Overall migration study of chitosan-encapsulated ZnO-based ESO bionanocomposite with synergistic antimicrobial activity for packaging purpose. Chem. Select (2022)
52. Pourjavadi, A., Farhadpour, B., Seidi, F.: Synthesis and investigation of swelling behavior of new agar based superabsorbent hydrogel as a candidate for agrochemical delivery. J. Polym. Res. (2009)
53. Mihaila, S.M., Gaharwar, A.K., Reis, R.L., Marques, A.P., Gomes, M.E., Khademhosseini, A.: Photocrosslinkable kappa-carrageenan hydrogels for tissue engineering applications. Adv. Healthc. Mater. (2013)
54. Ding, X., Li, L., Liu, P.S., Zhang, J., Zhou, N.L., Lu, S., Wei, S.H., Shen, J.: The preparation and properties of dextrin-graft-acrylic acid/montmorillonite superabsorbent nanocomposite. Polym. Composit. (2009)
55. Nnadi, F., Brave, C.: Environmentally friendly superabsorbent polymers for water conservation in agricultural lands. J. Soil Sci. Environ. Manage. (2011)
56. Coutinho, D.F., Sant, S.V., Shin, H., Oliveira, J.T., Gomes, M.E., Neves, N.M., Khademhosseini, A., Reis, R.L.: Modified gellan gum hydrogels with tunable physical and mechanical properties. Biomaterials (2010)
57. Jayakumar, R., Nagahama, H., Furuike, T., Tamura, H.: Synthesis of phosphorylated chitosan by novel method and its characterization. Int. J. Biol. Macromol. (2008)

58. Rinaudo, M., Le Dung, P., Gey, C., Milas, M.: Substituent distribution on O, N-carboxymethylchitosans by 1H and 13C NMR. Int. J. Biol. Macromol. (1992)
59. Chen, Y., Zhang, Y., Wang, F., Meng, W., Yang, X., Li, P., Jiang, J., Tan, H., Zheng, Y.: Preparation of porous carboxymethyl chitosan grafted poly (acrylic acid) superabsorbent by solvent precipitation and its application as a hemostatic wound dressing. Mater. Sci. Eng. C (2016)
60. Mi, F.L., Wu, S.J., Lin, F.M.: Adsorption of copper (II) ions by a chitosan–oxalate complex biosorbent. Int. J. Biol. Macromol. (2015)
61. Percival, E.: The polysaccharides of green, red and brown seaweeds: their basic structure, biosynthesis and function. Brit. Phycol. J. (1979)
62. Rinaudo, M.: Main properties and current applications of some polysaccharides as biomaterials. Polym. Int. (2008)
63. Rinaudo, M.: Biomaterials based on a natural polysaccharide: alginate (2014)
64. Lee, K.Y., Mooney, D.J.: Alginate: properties and biomedical applications. Prog. Polym. Sci. (2012)
65. Comaposada, J., Gou, P., Marcos, B., Arnau, J.: Physical properties of sodium alginate solutions and edible wet calcium alginate coatings. LWT-Food Sci. Technol. (2015)
66. Zhang, L., Liu, T., Chen, N., Jia, Y., Cai, R., Theis, W., Yang, X., Xia, Y., Yang, D., Yao, X.: Scalable and controllable synthesis of atomic metal electro catalysts assisted by an egg-box in alginate. J. Mater. Chem. A (2018)
67. Nallamuthu, N., Braden, M., Patel, M.P.: Dimensional changes of alginate dental impression materials. J. Mater. Sci.—Mater. Med. (2006)
68. Jin, S.G., Yousaf, A.M., Kim, K.S., Kim, D.W., Kim, D.S., Kim, J.K., Yong, C.S., Youn, Y.S., Kim, J.O., Choi, H.G.: Influence of hydrophilic polymers on functional properties and wound healing efficacy of hydrocolloid based wound dressings. Int. J. Pharm. (2016)
69. Cui, Z., Zhang, Y., Zhang, J., Kong, H., Tang, X., Pan, L., Xia, K., Aldalbahi, A., Li, A., Tai, R.: Sodium alginate-functionalized nanodiamonds as sustained chemotherapeutic drug-release vectors. Carbon (2016)
70. Dhivya, S., Padma, V.V., Santhini, E.: Wound dressings–a review. BioMedicine (2015)
71. Sirvio, J.A., Kolehmainen, A., Liimatainen, H., Niinimaki, J., Hormi, O.E.O.: Biocomposite cellulose-alginate films: promising packaging materials. Food Chem. (2014)
72. Le Corre, D., Bras, J., Dufresne, A.: Starch nanoparticles: a review. Biomacromolecules (2010)
73. Zou, W., Yu, L., Liu, X., Chen, L., Zhang, X., Qiao, D., Zhang, R.: Effects of amylose/amylopectin ratio on starch-based superabsorbent polymers. Carbohydr. Polym. (2012)
74. Agama, A.E., Flores Silva, P.C., Bello Perez, L.A.: Cereal Starch Production for Food Applications. Elsevier, Starches for Food Application (2019)

Theory of Superabsorbent Polymers

Sulena Pradhan and Sukanya Pradhan

Abstract Superabsorbents are a class of functional macromolecules that have hydrophilic polyhydroxy and slightly crosslinked three-dimensional (3D) network structures. Due to their exceptional biocompatibility, biodegradability, low toxicity, and natural metabolism in physiological systems, biopolymer-based superabsorbent hydrogels hold a lot of potential. Tridimensional crosslinked polymeric substances known as superabsorbents are able to quickly absorb water, physical fluids, and biological fluids up to multiple times their own weight without dissolving when in contact with water. The general theoretical elements of superabsorbents are covered in this chapter. Additionally, the factors determining the absorption capacities are also discussed.

1 Introduction

Superabsorbent polymers are a type of hydrogels that have an absorption capacity of 500–1500 g/g of water, whereas common hydrogels have an absorption capacity of no more than 1000 g/g. Generally speaking, superabsorbent polymers can be divided into several categories based on different factors: they can be divided into two categories based on the mechanism of water absorption: chemical and physical absorption; they can also be divided into about six categories: starch; cellulose such as cardiomyopathy cellulose (CMC)-based SAPs; protein; synthetic of polymers such as the copolymers of acrylic acid and acrylamide; chitosan, and finally the blends and composite such as organic–inorganic hybrid materials-based superabsorbent polymers [1–3].

It can also be divided into two categories based on the crosslinking process: self-crosslinking SAPs like polyacrylates and polyacrylamides that form networks through the addition of a crosslinker like poly(acrylic acid) crosslinked by metal

S. Pradhan
Larodan AB, Karolinska Institute, Solna, Sweden

S. Pradhan (✉)
SARP-LARPM, CIPET, Bhubaneswar, Odisha, India
e-mail: pradhan.sukanya2@gmail.com

S. Pradhan and S. Mohanty (eds.), *Bio-based Superabsorbents*, Engineering Materials,
https://doi.org/10.1007/978-981-99-3094-4_4

ions or starch/acrylate crosslinked by N,N′-methylenebisacrylamide. Few companies develop SAPs with biocompatibility and biodegradability qualities that are made from renewable materials [4].

In general, SAPs are used for a wide variety of purposes [5] in addition to sanitary ones, including agriculture (soil conditioning and water reserving) and pharmaceutics (controlled drug delivery to the gastrointestinal tract).

2 Absorption Versus Superabsorption

According to the two primary types of water absorption, chemical and physical, the hygroscopic substance is often divided into two categories. Chemical absorbers, such as metal hydrides, capture water by chemically changing their nature. Physical absorbers imbibe water via the following approaches; reversible changes of their crystal structure (e.g., silica gel and anhydrous inorganic salts); physical entrapment of water via capillary forces in their macro-porous structures (e.g., soft polyurethane sponge); a combination of the mechanism (ii) and hydration of functional group (e.g., tissues paper); the mechanism which maybe anticipated by combination of mechanism of (ii) and (iii) and essentially dissolution and thermodynamically favoured expansion of the macro-molecular chain limited by Silica gel and anhydrous inorganic salts are examples of physical absorbers that imbibe water. Other examples include soft polyurethane sponges that physically trap water via capillary forces.

The latter category includes superabsorbent polymer (SAP) materials, which are organic materials with a high capacity for water absorption. In contrast to common hydrogels, which have an absorption capacity of no more than 100% (1 g/g), SAPs as hydrogels can absorb and retain extraordinarily large amounts of water or aqueous solution relative to their own mass. These ultra-high absorbing materials can imbibe deionized water up to 1,000–100,000% (10–1000 g/g) [6]. Figure 1 provides schematic visual representations of a crosslinked hydrogel network. Swollen structure after immersion in water (adapted from [7]).

3 Swelling Mechanism

The design and fabrication of a SAP material involve two processes: swelling and crosslinking. Flory was the first to investigate the mechanism underlying the swelling behaviour of hydrophilic polymers in ionic and nonionic networks and discovered that the water absorption capacity of SAPs was inversely associated with the ionic strength of the solution being absorbed [8, 9].

Further research revealed that the pH of the liquid affects how SAPs swell. For instance, when the pH was increased from 2 to 6, the swelling capacity for a polyacrylamide-sodium allylsulfonate-sodium acrylate system increased from 70 to 450 g/g; however, when the pH was increased further to 13, it reduced to 180 g/

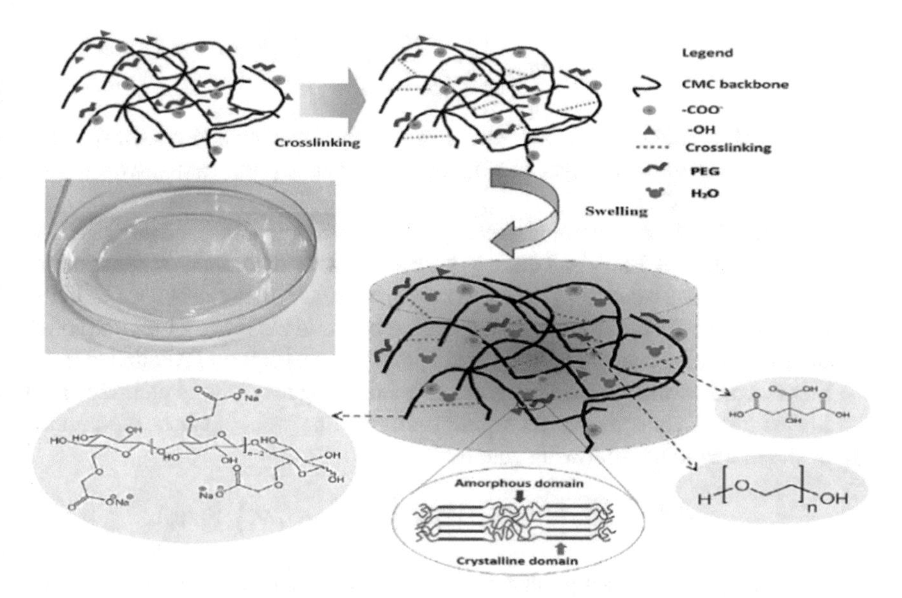

Fig. 1 Schematic visual representations of a crosslinked hydrogel network. Swollen structure after immersion in water

g [10]. The temperature has a considerable impact on SAPs' ability to expand, with higher temperatures typically resulting in greater swelling capacity. The swelling behaviour of these SAPs is predominantly affected by the following parameters:

(1) Nature of interaction: The swelling properties of ionic hydrogels are primarily influenced by four forces: interactions between the polymer and the solvent, ionic interactions, elastic interactions, and electrostatic repulsion. Because mobile ions like K^+ and Na^+ are present in the SAP structure, ionic contact is the most significant of these forces.

Chemical structure and hydrodynamic volume (HDV), or the volume occupied by the solvated chain, are the two factors that can best be used to forecast how polymers would behave in solutions. Simply put, this chain can be thought of as freely joined and having an end-to-end distance of 'r'. The following shows that the latter parameter can be connected to the quantity of bonds, n, and the length of each bond, 'l'.

$$\langle r^2 \rangle = nl^2 \tag{1}$$

But, by accounting for the valence bond angle, θ, and conformational angle φ, a more realistic character of this typical chain is obtained. In order to impose directionality on the chain, these two exhibit restriction or stiffening characteristics as represented in the following equation:

$$\langle r^2 \rangle = nl^2 \left(\frac{1 - \cos\theta}{1 + \cos\theta} \right) \left(\frac{1 + \overline{\cos\varphi}}{1 - \overline{\cos\varphi}} \right) \quad (2)$$

In SAP, a balance is reached between naturally occurring hydrophilic polymer segment dissolution and thermodynamically favoured polymer chain expansion that is crosslinking.

Using the Flory–Huggins model of the thermodynamics of polymer and solvent mixtures, Flory and colleagues provided the theoretical foundation for the equilibrium balance between the absorption and retention of water under pressure [11]. By incorporating rubber elasticity into the Flory–Huggins model, the Flory–Rehner swelling hypothesis predicted that lightly crosslinked polymers would absorb water more readily. By including the osmotic pressure effect in the Flory–Rehner theory, the equilibrium swelling ratio, Q, can be represented as illustrated in the following equation:

$$Q^{5/3} = \left\{\left[i/\left(2V_u S^{*1/2}\right)\right]2 + (1/2 - \chi_1)/V_1\right\}(V_0/v_e)$$

where i/V_u is fixed charge density in the polymer chain, S^* is ionic strength of solvent, V_1 is solvent molar volume, χ is an empirically derived interaction parameter expressing polymer–solvent affinity, v_e is the number of polymer chains comprising the network structure, and V_0 is the polymer volume before swelling (v_e/V_0 is the density of crosslinking).

In conclusion, the following elements affect a SAP's capacity:

1. Crosslinking density
2. The compatibility of a polymer with a solvent
3. Osmotic pressure brought on by counterions in polyelectrolytes

(2) Polarity: Typically, the amount of polar groups and the quantity of water molecules adsorbed are inversely correlated for natural polymers. The greater the number of hydrophilic groups, the greater the water absorption capacity.
(3) Elastic retroactive force: As the retroactive force, which develops as the chains between the crosslinks get longer, balances out the swelling force at some point, the swelling tendency will cease [12].
(4) Degree of crosslinking: The polymeric chain length is altered by the type of crosslinker used. Longer chains contain greater network space and so increase swelling, whereas smaller chains have more polymer ends and do not contribute to water absorption.
(5) Solvent properties: The concentration of the solvent has an influence according to Fick's law of diffusion because swelling involves the diffusion of the solvent molecules within the gel network structure.
(6) Morphological properties (porosity, particle size): the smaller the average grain size, the larger the swelling.
(7) Conditions of the swelling environment such as pH, temperature, ionic strength as well as the counter ion and its valance.

4 Crosslinking

Bulk

There are primarily two methods for crosslinking polymer chains (Fig. 2). The first involves small-scale free-radical copolymerization using polyvinyl comonomers (Fig. 2a). The most popular technique for synthesizing crosslinked polymers is by far the usage of polyunsaturated comonomers (acrylates). In this instance, chain expansion and crosslinking happen simultaneously. The most popular polyunsaturated crosslinkers are diacrylates or bisacrylamides, however, trifunctional acrylates are also occasionally utilized.

The most frequently utilized polyunsaturated crosslinkers are diacrylates or bisacrylamides, however, trifunctional acrylates and di- or triallyl compounds are also occasionally used. The second method uses suitable polyfunctional chemicals to initiate a nucleophile or condensation reaction between the carboxylate and carboxylic pendant groups of the linear polymer, respectively (Fig. 2b); poly (epoxides), halo-epoxides, and polyols are the most prominent. Without the need for crosslinker molecules, spontaneous chemical crosslinking or auto-crosslinking takes place by radical combinations or intermolecular interactions involving chemical functionalities that are already present in the polymer chains as pendant groups, such

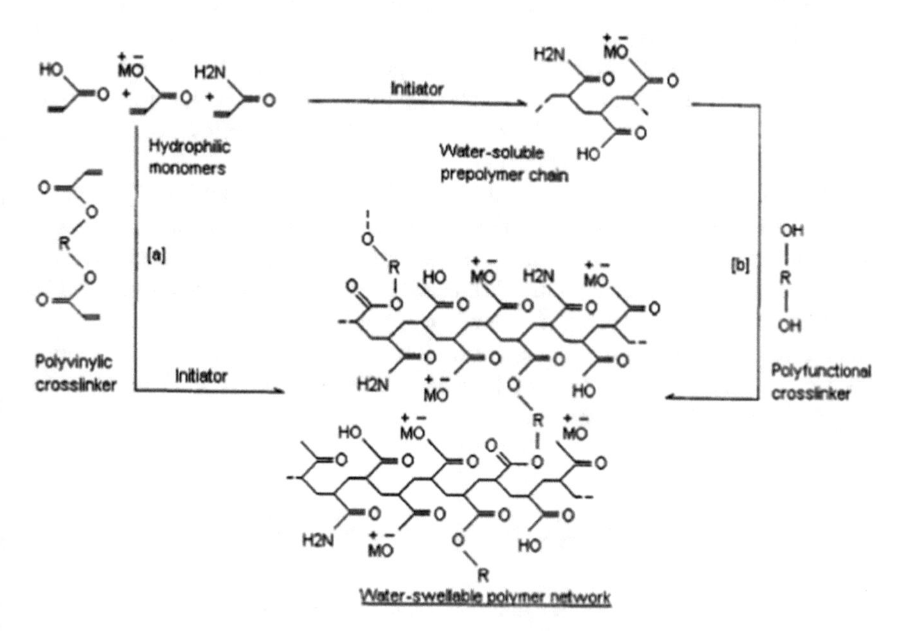

Fig. 2 Two essential approaches of crosslinking (adapted from [11]). **a** Simultaneous polymerization and crosslinking via radical mechanism. **b** Crosslinking after polymerization via functional groups condensation

as carboxyls and hydroxyls. Covalent crosslinks are synthesized by these crosslinking processes.

Treatment of polymerization mixtures with multivalent metallic cation solutions is another method for creating crosslinks. Ionic bridges between the carboxylate groups of two or more distinct poly(acrylate) chains can be created by multivalent metal ions.

After polymerization, zinc, calcium, strontium or barium acetates are utilized for the modification process. Crystalline domains (hard segments), similar to thermoplastic elastomers, serve as crosslinks. It appears that poly(vinyl alcohol) crystalline regions make block copolymers of an acrylic monomer and a cleavable monomer like vinyl alcohol insoluble. The polymer chains can be kept from dissolving by the hydrophobic interaction of pendant long chain hydrocarbon groups made by copolymerizing an acrylic monomer with a small amount of an oil-soluble comonomer, such as lauryl methacrylate. Certain absorbents may produce gel through a process known as hydrophilic association via hydrogen bonding.

Crosslinking at Surface

In fact, when water approaches a water-absorbent resin that has been crosslinked using any of the methods mentioned above, swelling first happens at the surface of the particles, delaying further water penetration into the core or towards the central sections of the particles. The particles eventually cling together and form agglomerates, which prevent water from diffusing towards the particles inside the globules. A low absorption rate is, therefore, achieved. The phenomena have also been referred to as "gel-blocking" and "fish-eyes" formation. Further, making the gel surface structure stiffer is one way to avoid these difficulties, hence the efficiency of surface crosslinking was found to be prevalent.

It is also possible to use electrophilic polysubstituted compounds or hydrophilic associations to cause surface crosslinking in superabsorbent polymers that have already undergone surface crosslinking and were recovered from the reaction system. Superabsorbents made from crosslinked acrylic can also be surface crosslinked by immersing the particles in glycerin or another polyol. Surface crosslinking also results via adding peroxide to the polymer slurry at the conclusion of the polymerization process and heating the mixture at a temperature that is 50–100 °C higher than the peroxide's degradation temperature.

5 Some Important Properties of Superabsorbents

The nature of the polymer and the additives have a great impact on swelling characteristics such as swelling capacity, swelling rate, reswellability, and salt sensitivity in terms of water absorption capacity. Here is a discussion of additional qualities that are affected by these SAPs composite composition.

5.1 Swelling Capacity

A superabsorbent polymer's swelling capacity can be altered by altering the type and concentration of the crosslinker, monomer, initiator, as well as the reaction temperature. The swelling/crosslinker concentration power law behaviour demonstrates that among these, the crosslinker concentration has the most significant impact.

Particle size, porosity, drying method, and polymer network structure are some of the variables that affect the swelling rate. Particle size, specifically swelling kinetics, is crucial to superabsorbent characteristics. Appropriate particle size is required for a particular application in order to obtain desired qualities. For instance, a hygiene SAP must be greater in size than an agriculture SAP. The more surface area and larger voids leads the swelling particles to swell more and improve the soils fertility. A sanitary superabsorbent, on the other hand, must absorb urine or blood as rapidly as possible, enabling the use of smaller particles.

5.2 Reswellability

Agricultural superabsorbents must have the ability to reswell (reswellability), as these tiny water reservoirs must maintain their water absorption capacity for years and through innumerable swelling–deswelling cycles.

5.3 Saline Sensitivity

It is well known that the salinity (ionic strength) of the swelling medium has a significant impact on the swelling capacity of ionic hydrogels and SAPs. The swelling capacity of the hydrogel is considerably reduced by increasing ionic strength, decreased osmotic pressure, and salt-polymer complexation, to the point where the hydrogel precipitates at high salt concentration. If the salts have multivalent ions, the impact of the salt is more pronounced.

The SAPs in saline solutions commonly exhibit on-and-off swelling behaviour. On–off swelling behaviour is adversely affected by multivalent cations. Using FTIR research, the impact of multivalent cations on the SAP structure can be monitored. Because there is no interaction between the salt and the carboxylate group when a SAP is submerged in NaCl solution, its carbonyl group remains intact. Due to the development of a compound between the carboxylate group and cations like Mg^{2+} or Al^{3+} in multivalent salt solutions, the carbonyl peak moves to a lower wave number [13].

5.4 Residual Monomer

Each hydrogel product's residual monomer content will vary depending on the parameters of the reaction, the kind and concentration of the initiator, the monomer, the reaction temperature, and other factors [5]. As an illustration, it was discovered that the rate at which the initiator dissociates can significantly alter the concentration of the residual monomer [12].

6 Conclusion

SAPs are polymers that are just moderately crosslinked and have a balance between naturally occurring dissolution and thermodynamically advantageous expansion of polymer chains controlled by crosslinking that allows them to quickly absorb and hold water. SAPs have been developed for usage in a variety of industries, from hygiene products to civil engineering, construction, and agriculture, due to their special capacity to absorb and retain water and the ability to tune other qualities. In the future, SAPs are anticipated to be crucial as intelligent materials (such as actuators, drug delivery systems, and artificial organs).

References

1. Lacoste, C., Lopez-Cuesta, J.M., Bergeret, A.: Development of a biobased superabsorbent polymer from recycled cellulose for diapers applications. Eur. Polymer J. **116**, 38–44 (2019)
2. Chen, J., Wu, J., Raffa, P., Picchioni, F., Koning, C.E.: Superabsorbent Polymers: from long-established, microplastics generating systems, to sustainable, biodegradable and future proof alternatives. Prog. Polym. Sci. **125**, 101475 (2022)
3. Fouilloux, H., Thomas, C.M.: Production and polymerization of biobased acrylates and analogs. Macromol. Rapid Commun. **42**(3), 2000530 (2021)
4. Bio-based superabsorbent polymers made from renewable resources. www.adm.com/en-US/products/industrial/superabsorbents/Pages/default.aspx
5. Zohourian, M.M., Kabiri, K.: Superabsorbent polymer materials: a review (2008)
6. Buchholz, F.L., Graham, A.T.: Modern Superabsorbent Polymer Technology. Wiley, New York, NY (1998)
7. Capanema, N.S., Mansur, A.A., de Jesus, A.C., Carvalho, S.M., de Oliveira, L.C., Mansur, H.S.: Superabsorbent crosslinked carboxymethyl cellulose-PEG hydrogels for potential wound dressing applications. Int. J. Biol. Macromol. **106**, 1218–1234 (2018)
8. Ravve, A., Ravve, A.: Physical properties and physical chemistry of polymers. In: Principles of Polymer Chemistry, pp. 17–67 (2012)
9. Yao, K.J., Zhou, W.J.: Synthesis and water absorbency of the copolymer of acrylamide with anionic monomers. J. Appl. Polym. Sci. **53**(11), 1533–1538 (1994)
10. Kang, H., Xie, J.: Effect of concentration and pH of solutions on the absorbency of polyacrylate superabsorbents. J. Appl. Polym. Sci. **88**(2), 494–499 (2003)
11. Omidian, H.: Improved Superabsorbent Polymers. Doctoral dissertation (1997)

12. Kabiri, K., Zohuriaan-Mehr, M.J., Bouhendi, H., Jamshidi, A., Ahmad-Khanbeigi, F.: Residual monomer in superabsorbent polymers: effects of the initiating system. J. Appl. Polym. Sci. **114**(4), 2533–2540 (2009)
13. Wang, W., Wang, A.: Preparation, swelling, and stimuli-responsive characteristics of superabsorbent nanocomposites based on carboxymethyl cellulose and rectorite. Polym. Adv. Technol. **22**(12), 1602–1611 (2011)

Chitin/Chitosan Based Superabsorbent Polymers

Swarnalata Sahoo

Abstract Chitin/chitosan biopolymers having significant structural capability for mechanical and chemical modifications to produce innovative properties, functions, and numerous applications particularly in the field of agriculture, biomedical, hygiene, and water treatment. These chitin/chitosan biopolymers are taken into consideration as an effectual polymeric material due to its biocompatibility, non-toxicity, antimicrobial, and biodegradability property. These biopolymers act as an absorbent polymer primarily derived from seafood waste based materials such as shrimp, waste crab, lobster shells, etc. Superabsorbent polymers (SPs) are the polymers that are able to soaking up water up to 1500 gm of water per gram of SP through hydrogen bonding with molecules of water. These polymers are generally cross-linked with polyelectrolytes which produce a hydrogel when contact with aqueous solutions. The development of novel superabsorbent polymeric materials through green technology and physical approaches or from raw based chitin/chitosan is a great pathway and beneficial to the "biobased development material" in industry. This chapter focuses on the current progress in chitin based superabsorbent polymer, preparation method approaches and preparation conditions, specific properties of SPs such as water absorbance properties, biodegradability, antibacterial, biocompatibility properties, etc. Further, a versatile application of superabsorbent polymers with short section of future scopes and demanding challenges has also been elaborated.

Keywords Biopolymer · Composite · Superabsorbent · Biodegradability

S. Sahoo (✉)
CIPET: IPT, BBSR, Bhubaneswar 751024, Odisha, India
e-mail: sahoo.swarnalata@gmail.com
URL: https://scholar.google.com/scholar?hl=en&as_sdt=0%2C5&q=swarnalata+sahoo&oq=swarnalat

S. Pradhan and S. Mohanty (eds.), *Bio-based Superabsorbents*, Engineering Materials,
https://doi.org/10.1007/978-981-99-3094-4_5

1 Introduction

1.1 Super Absorbent Polymer

Superabsorbent polymer materials (SPs) are the polymers that are constituted by water soluble building blocks in a cross-linked networks structure. These SPs are primarily made of ionic monomers with low cross-linking densities. Low cross-linking density materials results larger liquid uptake capacity i.e. up to 1000 times greater of their own weight. While in hydro gels, the polymer network structure can absorb and maintain aqueous solution up to 10 times greater to own weight [1–4]. Hence, there is a clear distinction between SPs and hydrogels [5–12]. The most important characteristic of SP is to show large water absorption capacity relative to the mass of the sample. High water absorption capacity in SPs is due to the produced osmotic pressure in water which forces water molecule in to polymer matrix due to its inside high ionic concentration as compared to its surrounding solution [13–15]. This is because, the occurrence of hydrophilic and charged moieties on to its ionic monomers induce hydrogen bonding and attract water by means of combination of charged group and polar moieties in SPs. The swelling capacity is directly proportional to the amount of induced ionic and polar groups [16, 17]. Hence, this affinity makes the super absorbent polymer useful specially in the area of personal hygienic products, medicine, agriculture, industrial absorbent, agriculture, cosmetic, etc. due to its exceptional characteristics such as biodegradability, non toxic, renewability, biocompatibility, biocompatibility, etc. However, the higher in production price and lower in gel strength make the superabsorbent polymers to restrict their application. Therefore, to overcome these issues, incorporation of low cost based inorganic compounds can be utilized as an inorganic filler within the polymer matrix increases its mechanical strength with stiffness.

1.2 Chitin/Chitosan Based Super Absorbent Polymer

Mostly lots of shrimps and prawns are found in Bangladesh as it is crisscrossed with lots of rivers and has broad coastal lines which generally stretch about 750 kms. These shrimps and prawn shells are exports to various region of the world such as Europe and North America. Due to no utilization of shells causes a threat to the environment as an unprocessed like garbage. More efforts have been given to extract polymer from this garbage. Generally, 'chitin' derived from Greek word 'chiton' which means a coat of mail [18]. The structure of cellulose and chitin $(C_8H_{13}O_5N)n$ is almost similar but the monomer unit 2-acetamido-2-deoxy-β-D-glucose are attached to each other through linkages of $\beta(1 \rightarrow 4)$. It can be considered as similar to the structure of cellulose whereas the hydroxyl group of chitin was replaced by a group of acetamido at a position of C-2. Chitin polymers are extracted from the shrimps and prawn shell garbage and then it is transformed in to its deacetylated derivative chitosan polymer

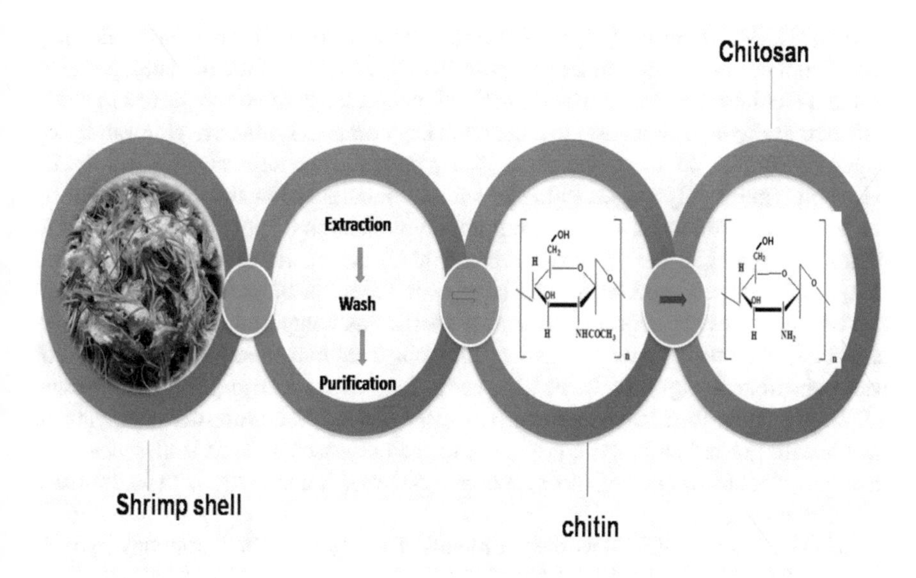

Fig. 1 Extraction process of chitin and chitosan from shrimp shell

through some further treatment represented in Fig. 1 [19]. Considering to the properties, chitin polymer is non reactive to chemical and extremely insoluble substance and chitosan is soluble in acidic solution. Both containing high percentage of nitrogen i.e. approx. 6.89% as compared to the cellulose derived synthetic substitute which contains only 1.25% of nitrogen. Most of the polysachharides occurred naturally are such as agar, cellulose, agarose, dextran, alginic, pectin, alginic acid, carrageenan, etc. generally neutral or some acidic in nature, but chitosan is highly basic in nature based polysaccharide.

Chitin was first invented in year 1811 by the chemist Henri Braconnot. Annually the estimated natural production rate of chitin is approx. 1011 ton. These biodegradable polymers can be easily refined into a variety of products such as nanofibers, hydrogels, membranes, nanofibres, nanoparticles, etc. [20]. Moreover, in dressings and to stop bleeding of mucous cell, chitosan based sponges are used. Similarly in contact lenses, in dialysis and in dressings, chitin based polymeric film and membranes are used [21–31]. In addition to this, a special focus has been given to nanocomposites due to its intercalation properties and smaller particle size of nanomaterial. Among the nanofiller, mineral powders are the one who make the polymer as superabsorbent polymer composite as there is a strong interaction between mineral powder and the reactive sites of natural polymer. Synthetic polymer or natural polymer based superabsorbent polymer composites already have been developed by some of authors [7–10]. Among all the superabsorbent polymer composite, chitin/chitosan based superabsorbent polymer composite plays a vital role due to its excellent properties such as biocompatibility and biodegradability.

In 2005, Maeda et al. [32] started research work to synthesize and to develop recent superabsorbent polymer composites with the incorporation of kaolin particles inside the chitosan polymer. They found that the presence of amino groups in chitin and their ability to protonation was the main key points to synthesize chitosan-based superabsorbrnent polymer composite. They prepared the superabsorbent polymer based on natural polymer i.e. chitosan with the incorporation of inorganic filler by the polymerization technique i.e. graft copolymerization of poly acrylic acid on to chitosan polymer in the presence of crosslinking agent. After that, they analysed that, the developed super absorbent composite based on biopolymer is cost effective and can provide good mechanical properties as compared to synthetic superabsorbent polymer composite. Further, they observed that it had a larger degree of water absorbance capacity. In FTIR spectra, they found a absorption peak around $1722\,cm^{-1}$ due to formation of ester group by the replacement group of hydroxyl from kaolin with grafted carboxylic groups in to the polysaccharide. It is also observed that the water absorbency of super absorbent polymer is decreased with the increase in kaolin content.

In 2006, Prasad et al. [33] worked on blends of agar grafted PVP (polyvinylpyrrolidone) with k-carrageenan-graft-PVP which capable of forming of hydrogels product. They initiated by carring out the reaction with the process of microwave irradiation in the presence of a initiator such as water-soluble and potassium persulfate. They studied the thermal stability and structural characteristics of the grafted blends by using thermogravimetric, Fourier transform infrared and 13C-NMR analyses. They revealed that the crystallinity increases in the products as compared the control PVP as well as polysaccharides with proposing a possible mechanism for the crosslinking of polymer PVP to agar and then with k-carrageenan. They concluded that the developed graft blends may be of potential utilization in tissue engineering, in biomedicine, as a water retainer in agriculture, in pharmaceuticals, in microbiology and as hydrogels. They also observed that the hydrogels showed higher water absorbency capacity despite of weaker gel strength as compared with respective polysaccharides.

In 2008, Liu et al. [34] synthesized sustainable super absorbent polymer using chemically modified pulverized wheat straw (CMPWS) and acrylic acid (AA) in presence of initiator and cross linker. They investigated the factors which are affecting the absorbency properties of super absorbent materials such as initiator, cross linker amount, weight ratio between AA to CMPWS, temperature, etc. They also studied about the absorbency capacity and found that in distilled water, the maximum absorbency is 417 g/g and in 0.9 wt% of NaCl solution, the maximum absorbency is 45 g/g. Hence, they observed that, the cost of production was significantly reduced that of cross linked poly acrylic-co-acrylamide super absorbent polymer.

In 2015, Li et al. [35] synthesized semi interpenetrating polymer network (Semi IPNs) hydrogel based on a wheat straw and slow-release fertilizers such as phosphorus and nitrogen. They have utilized dipotassium hydrogen phosphate and urea to supply the phosphorus and nitrogen as nutrients, respectively. They examined especially on the swelling properties of material at water and salt solution, pH, ionic strength, etc. and observed that the developed material can hold the water at pH level 6–9 with the effect of cations in the order as $Na+ > K+ > Ca^2+$ on fertilizer release and swelling properties. They found that in distilled water, the water absorbency is 198.5 g/g and in 0.9 wt% of sodium chloride solution, the water absorbency is 26.3 g/g. In addition to this, they also observed that better controlled release effect can be achieved due to the diffusion coefficient of the developed material.

Similarly, in 2015, Cheng et al. [36] synthesized hydrogel based on corn straw-co-AMPS-co-AA by means of removing the lignin from the corn straw and then modifying chemically using AMPS to sulfonated cellulose. They prepared in the presence of initiator and crosslinking agent i.e. potassium persulfate and N, N1-methylenebisacrylamide respectively. They observed the effects of parameters such as temperature, initiator, monomer neutralization degree, crosslinker, etc. on the swelling properties of the developed hydrogel. They also observed other properties such as salt resistance, water retention, recyclability, etc. and found that the optimum water absorption at 50 °C.

2 Chemical Structure and Properties of Chitin/Chitosan

Chitin is a long molecular chain of N-acetyl-glucosamine units joined together by poly(1–4)-β linkages as represented in Fig. 2. The chemical structure of chitin is resemblance with cellulose. The each sugar unit of its structure are rotated to an angle 180° with respect to each other. Generally, chitin has 3 different types of crystalline based allomorphs i.e. α-, β- and γ-forms. The most common form chitin is known as α-chitin which consists of two N,N′-diacetylchitobiose units forming two antiparallel chains and the adjacent molecular chains are in opposite direction which are held together by six hydrogen bond. The less common form of chitin is β-chitin which consists of a N,N′-diacetylchitobiose units forming parallel chains and the adjacent molecular chains are held together by three to five intermolecular

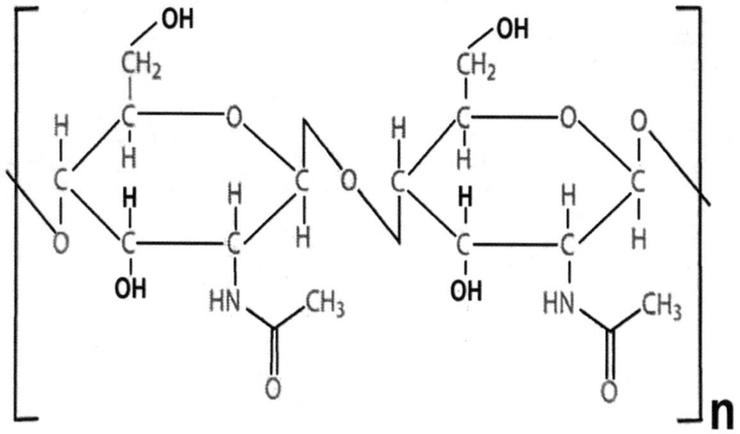

Fig. 2 Chemical structure of chitin

H bond. Hence, β-chitin allows more flexibility that α-chitin due to the parallel arrangement of polymeric chain in β-chitin [37]. The third allomorph of chitin is γ-Chitin which is mixed of parallel and anti parallel arrangement of polymer chain. Chitosan is formed from the *N*-deacetylation of the chitin polymer. The deacetylation process happens only under particular relatively conditions of hydrolysis, particularly when chitin is treated with concentrated aqueous solution of NaOH at a particular temperature of 383–413 K for the time period of 4–6 h. Generally in chitin, the degree of deacetylation is 0.8–0.9 which has never reaches to 1. The degree of acetylation of chitin indicates the occurrence of large amounts of 2-acetamido-2-deoxy-d-glucopyranose that contributes to the strong hydrogen bonding among the polymer chains of chitin [22]. The main source of chitin is shrimp and Crab shells due to the presence of high chitin amount and high annual production. The crab or lobster shell, chitin existing the form of nanofibers with diameters of around 3 nm [23, 24]. These chitin nanofibers are wrapped in protein to form the chitin-protein and subsequently embedded in minerals (mainly consisting of crystalline calcium carbonate and a small amount of calcium phosphate) to form the hard stratum corneum of lobster shell [37].

The process of formation of chitin from shrimp shell is described in Fig. 3 and as below:

1. Initially, shrimp shells were washed several times with sufficient potable water and then dried for a period of few minutes.
2. After that, it is again washed with cold water for the removal of acid.
3. Further, 4 ml of 1.25N HCL was added and kept overnight at room temperature. 1000 °C for the period of 5–6 h.
4. Similarly, 5 ml of 5% NaOH added and heated at temperature 70–75 °C for the period 2 h.
5. Then washed with water for removal of alkali and dried for 8 h at 60 °C.

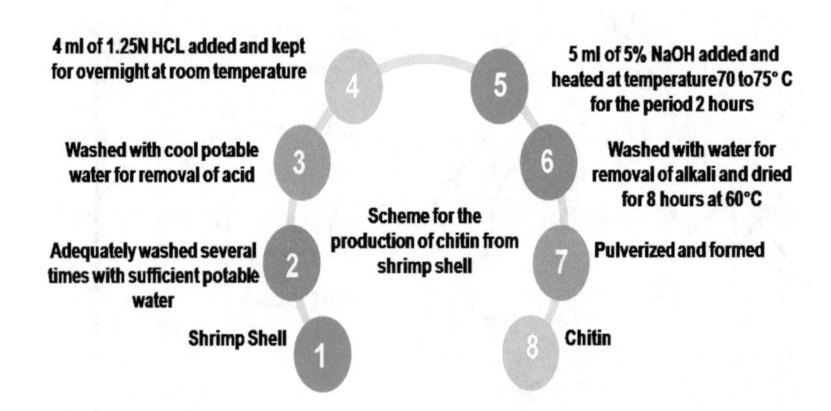

Fig. 3 Various steps for formation chitin from shrimp cell

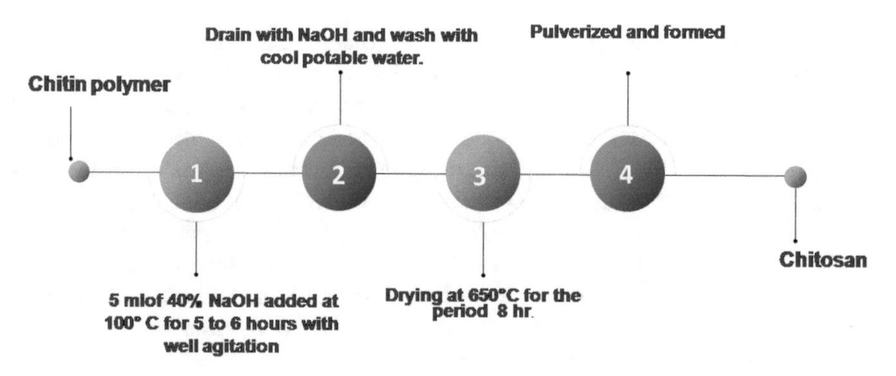

Fig. 4 Various steps for formation chitosan from chitin

6. Finally, the developed product was pulverized and formed as Chitin.

The process of formation of chitosan from chitin polymer is described in Fig. 4 and as below:

1. Initially, the chitin from shrimp shell is washed and then dried for a period of few minutes.
2. After that, 5 ml of 40% NaOH was added and heated at the temperature 1000 °C for the period of 5–6 h.
3. Further, NaOH solution was drained from the product and was washed with the help of potable water for many times and then dried for 8 h at 650 °C.
4. Finally, the developed product was pulverized by size and formed as Chitosan.

The unique properties of chitin and chitosan are to form polyoxysalts and have ability to produce films, non-toxicity, biodegradability with biocompatibility nature, adsorption properties, etc. Due to its excellent bioactive and biodegradability properties, there is a high demand for chitosan in the field especially in agriculture and pharmaceutical [38–41]. Lots of analytical techniques have been used to analyse the deacetylation using IR spectroscopy, ^{13}C solid state NMR, ^{1}H NMR spectroscopy, UV–vis spectrophotometry, thermal analysis, mechanical analysis e etc.

3 Solubility of Chitin and Chitosan Based Super Absorbent Polymer

The maximum amount of a substance of polymer which dissolves in a specified amount of solvent at a particular temperature is known as solubility. Hence, to choose appropriate solvent is most important to check the solubility of the polymeric material. The solubility of the polymeric material depends upon the molecular weight and molecular weight of the material depends upon the intrinsic velocity. The solubility parameter and cohesive energy of chitin is very high due to its semi-crystalline structure with wide-ranging hydrogen bonding. Therefore it is insoluble in all usual solvents [20, 42]. The solubility parameter value of chitin and chitosan was determined through group contribution method and the obtained values were compared with the values obtained from the Flory–Huggins model, surface tension and intrinsic viscosity. The solubility parameter value of chitin is more or less equal with the values determined from group contribution and from Flory–Huggins model [43]. The solubility parameter of chitin can be improved by further chemical treatment with strong aq. Hcl. Firstly, the solubility of chitin with respect to various solvents was carried out by Pillai et al. [44]. The solubility depends upon the molecular weight and intrinsic viscosity which can be calculated by using Mark Houwink equation (k).

According to Mark Houwink equation

$$\eta = KM^{\alpha}$$

where

η = Intrinsic viscosity.
K & α are the constants.
M = Molecular weight.

At low pH value, chitosan can easily produce quaternary nitrogen salt. So, chitosan can dissolve in some organic compounds such as formic acid, acitic acid, and lactic acid [18–20]. The solvent which is best for the chitosan is formic acid where solutions can formed in aqueous medium containing formic acid 0.2–100%.

4 Methodology for Super Absorbent Polymer

4.1 *Step 1—Synthesis Root or Chemical Cross Linking Method*

Based on the reaction mechanism, chemical cross-linking methods are of two types viz. (1) free radical inducing polymerization and (2) non free radical technique of polymerization. Free radical inducing polymerization is a typical technique and extensively approach for the synthesis of super absorbent based hydrogel. The recent studies explains about the superabsorbent chitosan hydrogel based on hydrophilic vinyl monomer, poly (acrylamide) and poly(acrylic acid). This is because, this free radical polymerization technique is highly convenient, efficient and effortlessly polymerized to produce high molecular weight based polymer [30]. It can also achieve through by non free radical technique of polymerisation (Fig. 5).

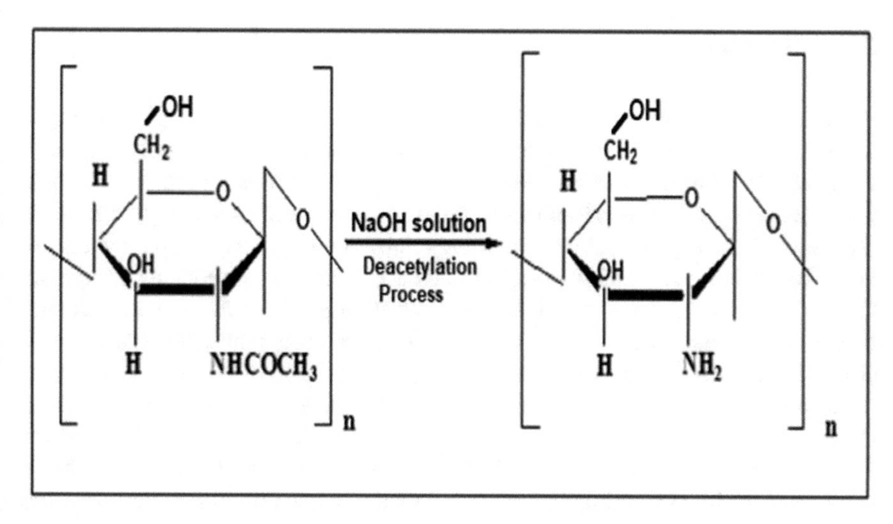

Fig. 5 Chemical cross linking method

4.2 *Step II—Physical Cross Linking Method*

It is known as Ionic cross linking method through which three dimensional based network polymer can be prepared without any non toxic chemical based agent. Whereas, the agent utilized in chemical cross linking method is toxic. Azuma et al. [21] worked on complex gel-beads in which the absorbency is up to 3675 g/g through the electrostatic based interactions between Al3+, –NH3+ and –COO– in a two stage physical cross linking process of chitosan and hydrolyzed polyacrylamide (HPAM). They observed that, the swelling ratio of super absorbent polymer grafted poly acrylic acid is much less than the hydrogel. This is because of the porous present in the surface of hydrogel. Similarly, GR et al. synthesized k-carrageenan with chitosan based hydrogel through electrostatic interactions without using any toxic agents. They observed that the unmodified chitosan have less water absorbency capacity as compared to modified one [22, 23].

5 Applications

Super absorbent polymer based on chitin/chitosan found a lot of application in the field of personal hygiene based products, agriculture, building, biomedical, water treatment, etc. Super absorbent polymer are invented mainly in companies of Japan such as Sumitomo Seika Chemicals, Nippon Shokubai, SDP Global, and companies of Japans such as Evonik, BASF, etc. These companies are produced super absorbent polymer approximately 70% of the Global market, though it is very much difficult to get information on the utilization of super absorbent polymer.

5.1 Biomedical

Since the superabsorbent material have their capacity to absorb and keep larger amount of fluids such as blood, urine, etc. Hence super absorbent polymer based materials are mainly used in the area of personal hygiene for example female napkins, disposal diapers, etc. and in biomedical fields for example drug delivery, controlled-release, various wound dressings, etc. Super absorbent polymer based diapers are generally sophisticated and well engineered based commercial products. Moreover, the baby diapers based on super absorbent production is about 80% of overall production [45]. It is also known as the safety of the superabsorbent materials. Marreco et al. [23] provided a stepwise systematic approach to produce a super absorbent from chitosan for the evaluation of safety of baby diapers and they reported about the antibacterial activity of chitosan based superabsorbent which can be used for female hygiene products and in baby diapers. After that Fan et al. [27] produced pH responsive for controlling the release of methotrexate drug for anticancer using chitosan with k-carrageenan. They observed that, the release of methotrexate at 5.3 pH is about 50% in 4 h where as at pH level 7.4 it releases about 68% in same 4 h. Moreover, Jayakumar et al. [24] developed a sustained release based superabsorbent materials to reduce the irritating effect produced from risedronate-sodium. Madhumathi et al. [25] reported chitosan-co-poly(AMPS) cross linked hydrogel in which molecular weight chitosan is very low for pH responsive. Further, Gopalan Nair and Dufresne [26] synthesized hydrogels composite for the discharge of theophylline and tinidazole. It was observed that the discharge of both drugs from the hydrogels showed closely fitting to the model of Korsmeyer–Peppas.

5.2 Agriculture

Two essential components of agriculture is fertilizer and water for the production of plant growth. However, most of the time it is not possible to utilize the fertilizer and water completely because of they are not environmentally friendly and not economical. Hence superabsorbent polymers based on polysaccharides with the incorporation of the fertilizer have gained more consideration, especially in the desert area. The antibacterial or the antimicrobial activity of chitin/chitosan polymer is only due to the molecular weight, pH of the solution, concentration, viscosity and degree of deacetylation [25, 26]. In horticulture, the chitin/chitosan has been incorporated in to soil to introduce as a nutrient and induces biotic and abiotic stress in various crops and helps in growth of the plant without affecting any soil natural properties. It has also the efficiency of combination with other industrial fertilizer to give plant-protection, plant growth simulation, post-harvest application, etc. [24].

Ifuku et al. [28] synthesized a three layered structure in which inner part coating is made of chitosan and core consisting of fertilizer. Further, the three layered structure coated with poly acrylic acid-co-acrylamide super absorbent polymer to manage fertilizer release and water absorbency. Further they developed a double coated multi-functional material which is controlled release of fertilizer. Similarly, Yusof et al. [29] synthesized tri layer coated fertilizer based hydrogel using chitosan and PVA containing nitrogen, potassium and phosphorous in situ suspension copolymerization reaction. Further, Prabaharan [30] developed a gellan gum based chitosan superabsorbent polymer to enhance the water absorbency ability of the soil. Correspondingly Jamnongkan et al. prepared fertilizer controlled release hydrogel which exhibited the equilibrium swelling ratio in the range 70–300% and increased the water retention ability of the soil.

5.3 Dressing

Generally the surgical-dressing consists of a layer of flesh colored tape placed upon the adherent based solution applied to the affected part of skin. The polymeric materials used to make surgical or wound-dressing are may be natural or synthetic in the form of films, hydrogels, sponges, etc. Extensively, wound-dressings are made from polysaccharides which are naturally occurring polymer [45, 46]. Further, the wound-dressings are various types such as primary wound-dressing, secondary wound-dressing, protective, bandages, adhesive made tapes, absorbents, etc. The basic purpose of this dressing is to provide temporary protective barrier, absorb the wound area drainage and to improve reepithelialisation by providing the necessary moisture absorption. Hence in this book chapter, a special attention has been given to biomaterial based dressing substitute for actively healing the wound process.

Prabaharan and Jayakumar [31] synthesized surgical fibrous dressing using chitosan based material by dissolving it in water in presence of appropriate acid to maintain the pH level which is about to 2–3% and also with the addition of gelatine dissolving in water. In this synthesis process, they kept the ratio between chitosan and gelatine is 3:1, 1:3 as well. Further, to decrease its stiffness, a small amount of plasticizer viz. glycerol/sorbitol was added to the mixture. After that, they casted the solution at room temperature through flat plate and produced a film which will be used as dressing material. Similarly, British textile technology [47] patented for making chitin and chitosan fibres which were obtained through the shell of micro-fungi not from the shell of shrimps.

5.4 Water Treatment

It is a process by which the quality of water is improved for particular end use of application. Various water treatments are constituted as sedimentation, flocculation, coagulation, disinfection, filtration, etc. Recently, the industry has given more focus about the water pollution by considering the environmental problems. Hence, it is a great challenge to get clean water based resources by developing new technique to filter the waste water. Hence, Maeda et al. [32] developed crosslinked polymer based on chitosan grafted polymaleic acid which is greater adsorbent uptake for mercury (Hg^{2+}) from few heavy metal ions such as Pb, Co, Cd, Hg, etc. present in the waste water. Similarly, Prasad et al. [33] developed chitosan based superabsorbent polymer composite with the incorporation of micro and nano bentonite material based clay. They analysed that the developed material is able to remove the methyl violet days and malachite-green from the waste water for water purification.

5.5 *Building*

In the current years, the superabsorbent polymers have been studied for concrete application. This is mainly concerned for the self healing of cracks created in concrete (11–15). In the case of concrete water plays the vital role for the hydration of cement and to control the deterioration which caused due to humidity. Similarly, water can also cause the early cracking of concrete due to the autogenous process of shrinkage. To overcome the above issues, superabsorbent polymer composites based on chitosan were used to avoid the early cracking of concrete. This is because; the structure created by concrete does not allow the water molecule from the outer surface to inside of the concrete during the early hardening process. Hence, the used cement needs extra water molecule to hydrate the concrete and make it harden. In order to overcome the hardening process of cement and to mitigate the cracks created in concrete, chitosan based superabsorbent composite have been developed as bio sourced materials [45, 46].

6 Demanding Challenges and Future Aspects

Every year efforts have been given to analyse and to prepare several type of chitin/chitosan based superabsorbent polymer and polymer composite using availability resource in order to balance the specific requirement such as social requirements, economical and technological point of view. To investigate the utilization and properties of several kind of fillers for using in it matrix in both research and industry, different compound strategy has been prepared for the unification of chitin/chitosan based superabsorbent polymer. It is also necessary to clarify the way in which the fillers affect the water absorbent capacity of the material. The advantages of incorporation of hybrid filler within polymer matrices instead of using single filler particles can lead to a number of applications because the incorporation of single filler is insufficient to achieve the desired property. As fillers are expensive, the hybrid filler incorporation in to the chitin/chitosan polymer to make superabsorbent polymer seems challenging and promising to research as well as industry. Further, for achieving both higher mechanical strength and higher water absorbency capacity, it is very much difficult to get. Hence, it is necessary to study the expansion properties of filler. Also, the inadequate knowledge for the mechanism of regeneration and dissolution hinders the various application of chitin. The high crystalline structure, high molecular weight and strong hydrogen bonding of chitin molecular chain shows a challenge for the preparation and processing of chitin.

7 Conclusion

Chitin and Chitosan are the biopolymers which have immense structural potential for mechanical and chemical modifications for the development of innovative functions, properties and applications especially in personal hygiene. Due to its huge amount of availability, the use of chitin based materials has been restricted due its insolubility properties. Several experiments on this have been reviewed for solving these issues. In addition to this, the presence of high viscosities of chitin in presence of certain solvent produces difficulties in processing. A detail descriptions on chitin/chitosan based superabsorbent polymers and polymer based composite on various potential applications such as agriculture, personal hygiene, biomedical, water treatment, etc. are reported. Several approaches were focused to utilize several kinds of super absorbent polymer/super absorbent polymer composite to keep balance on the mechanical properties of polymeric material along with its resistant to its chemical. It has also been reported the influence of incorporation of organic/inorganic fillers within the bio polymeric component. Although, incorporating fillers within a polymer matrix could increase the mechanical properties and higher water absorbency capacity of the developed material, still it is a big challenge to produce biodegradable super absorbent polymer composite by balancing both good mechanical strength and high water absorbency capacity with higher degree of substitution.

References

1. Buchhulz, F.L., Peppas, N.A.: Superabsorbent Polymer Science and Technology, ACS Symposium Series 573, American Chemical Society, Washington, DC (1994)
2. Buchholz, F.L., Graham, A.T.: Modern Superabsorbent Polymer Technology. Wiley, New York (1997)
3. Hennink, W.E., Van Nostrum C.F.: Adv. Drug. Deliv. Rev. **54**, 13 (2002)
4. Mignon, A., De Belie, N., Dubruel, P., Van Vlierberghe, S.: Superabsorbent polymers: a review on the characteristics and applications of synthetic, polysaccharide-based, semi-synthetic and 'smart' derivatives. Eur. Polymer J. **117**, 165–178 (2019)
5. Po, R.: J. Macromol. Sci-Rev. Macromol. Chem. Phys. **34C**, 607 (1994)
6. Peppas, L.B., Harland R.S.: Absorbent Polymer Technology. Elsevier (1990)
7. Bashari, A., Shirvan, A.R., Shakeri, M.: Cellulose-based hydrogels for personal care products. Polym. Adv. Technol. **29**(12) (2018)
8. Percival, S.L., McCarty, S.M.: Silver and alginates: role in wound healing and biofilm control. Adv. Wound Care **4**(7), 407–414 (2015)
9. Hoffman A.S.: In: Salamone J.C. (ed.) Polymeric materials encyclopedia, vol. 5. CRC Press, Boca Raton, FL (1996)

10. Mussatto, S.I., van Loosdrecht, M.: Cellulose: a key polymer for a greener, healthier, and bio-based future. Biofuel Res. J. **3**(4), 482 (2016)
11. Voisin, H., Bergström, L., Liu, P., Mathew, A.P.: Nanocellulose-based materials for water purification. Nanomaterials **7**(3), 57 (2017)
12. Huettermann, A., Orikiriza, L.J., Agaba, H.: Application of superabsorbent polymers for improving the ecological chemistry of degraded or polluted lands. CLEAN–Soil Air, Water **37**(7), 517–526 (2009)
13. Thakur, S., Govender, P.P., Mamo, M.A., Tamulevicius, S., Mishra, Y.K., Thakur, V.K.: (2017) Progress in lignin hydrogels and nanocomposites for water purification: future perspectives. Vacuum **146**, 342–355
14. Umachitra, G.: Disposable baby diaper—a threat to the health and environment. J. Environ. Sci. Eng. **54**(3):447–452 (2012)
15. La, Y.-H., McCloskey, B.D., Sooriyakumaran, R., Vora, A., Freeman, B., Nassar, M., Hedrick, J., Nelson, A., Allen, R.: Bifunctional hydrogel coatings for water purification membranes: improved fouling resistance and antimicrobial activity. J. Membr. Sci. **372**(1–2), 285–291 (2011)
16. Elieh-Ali-Komi, D., Hamblin, M.R.: Chitin and chitosan: production and application of versatile biomedical nanomaterials. Int. J. Adv. Res. **4**(3), 411 (2016)
17. Schröfl, C., Mechtcherine, V., Gorges, M.: Relation between the molecular structure and the efficiency of superabsorbent polymers (SAP) as concrete admixture to mitigate autogenous shrinkage. Cem. Concr. Res. **42**(6), 865–873 (2012)
18. Bidgoli, H., Zamani, A., Taherzadeh, M.J.: Effect of carboxymethylation conditions on the water-binding capacity of chitosan-based superabsorbents. Carbohyd. Res. **345**(18), 2683–2689 (2010)
19. Choudhury, N.A., Sampath, S., Shukla, A.K.: Hydrogel-polymer electrolytes for electrochemical capacitors: an overview. Energy Environ. Sci. **2**(1), 55–67 (2009)
20. Duan, B., Huang, Y., Lu, A., Zhang, L.: Recent advances in chitin based materials constructed via physical methods. Prog. Polym. Sci. **82**, 1–33 (2018)
21. Azuma, K., Ifuku, S., Osaki, T., Okamoto, Y., Minami, S.: Preparation and biomedical applications of chitin and chitosan nanofibers. J. Biomed. Nanotechnol. **10**(10), 2891–2920 (2014)
22. Yusof, N.L., Wee, A., Lim, L.Y., Khor, E.: Flexible chitin films as potential wound-dressing materials: wound model studies. J. Biomed. Mater. Res. Part A **66**(2), 224–232 (2003)
23. Marreco, P.R., da Luz, M.P., Genari, S.C., Moraes, A.M.: Effects of different sterilization methods on the morphology, mechanical properties, and cytotoxicity of chitosan membranes used as wound dressings. J. Biomed. Mater. Res. B Appl. Biomater. **71**(2), 268–277 (2004)
24. Jayakumar, R., New, N., Tokura, S., Tamura, H.: Sulfated chitin and chitosan as novel biomaterials. Int. J. Biol. Macromol. **40**(3), 175–181 (2007)
25. Madhumathi, K., Binulal, N.S., Nagahama, H., Tamura, H., Shalumon, K.T., Selvamurugan, N., et al.: Preparation and characterization of novel beta-chitin-hydroxyapatite composite membranes for tissue engineering applications. Int. J. Biol. Macromol. **44**(1), 1–5 (2009)
26. Gopalan Nair, K., Dufresne, A.: Crab shell chitin whisker reinforced natural rubber nanocomposites. Processing and swelling behavior. Biomacromolecules **4**(3), 657–665 (2003)
27. Fan, Y., Saito, T., Isogai, A.: Preparation of chitin nanofibers from squid pen beta-chitin by simple mechanical treatment under acid conditions. Biomacromol **9**(7), 1919–1923 (2008)
28. Ifuku, S., Nogi, M., Abe, K., Yoshioka, M., Morimoto, M., Saimoto, H., et al.: Preparation of chitinnanofibers with a uniform width as alpha-chitin from crab shells. Biomacromol **10**(6), 1584–1588 (2009)
29. Yusof, N.L., Lim, L.Y., Khor, E.: Preparation and characterization of chitin beads as a wound dressing precursor. J. Biomed. Mater. Res. **54**(1), 59–68 (2001)
30. Prabaharan, M.: Review paper: chitosan derivatives as promising materials for controlled drug-delivery. J. Biomater. Appl. **23**(1), 5–36 (2008)
31. Prabaharan, M., Jayakumar, R.: Chitosan-graft-beta-cyclodextrin scaffolds with controlled drug release capability for tissue engineering applications. Int. J. Biol. Macromol. **44**(4), 320–325 (2009)

32. Maeda, Y., Jayakumar, R., Nagahama, H., Furuike, T., Tamura, H.: Synthesis, characterization and bioactivity studies of novel beta-chitin scaffolds for tissue-engineering applications. Int. J. Biol. Macromol. **42**(5), 463–467 (2008). [PubMed: 18439672]
33. Prasad, K., Murakami, M.A., Kaneko, Y., Takada, A., Nakamura, Y., Kadokawa, J.I.: Weak gel of chitin with ionic liquid, 1-allyl-3-methylimidazolium bromide. Int. J. Biol. Macromol. **45**(3), 221–225 (2009)
34. Liu, Z., Miao, Y., Wang, Z., Yin, G.: Synthesis and characterization of a novel super-absorbent based on chemically modified pulverized wheat straw and acrylic acid. Carbohyd. Polym. **77**(1), 131–135 (2009)
35. Li, X., Li, Q., Xu, X., Su, Y., Yue, Q., Gao, B.: Characterization, swelling and slow-release properties of a new controlled release fertilizer based on wheat straw cellulose hydrogel. J. Taiwan Inst. Chem. Eng. **60**, 564–572 (2016)
36. Cheng, W.M., Hu, X.M., Wang, D.M., Liu, G.H.: Preparation and characteristics of corn straw-Co-AMPS-Co-AA superabsorbent hydrogel. Polymers **7**(11), 2431–2445 (2015)
37. Szosland, L.: Synthesis of highly substituted butyryl chitin in the presence of perchloric acid. J. Bioact. Compat. Polym. **11**, 61–71 (1996)
38. Pillai, C.K.S., Paul, W., Sharma, C.P.: Chitin and chitosan polymers: chemistry, solubility and fiber formation. Prog. Polym. Sci. **34**(7), 641–678 (2009)
39. Batista, R.A., Otoni, C.G., Espitia, P.J.: Fundamentals of chitosan-based hydrogels: elaboration and characterization techniques. Mater. Biomed. Eng. 61–81 (2019)
40. Zhai, M., Wu, G., Xu, L.: Radiation Processed Materials in Products from Polymers for Agricultural Applications in China. Chapter 7 (2014)
41. Islam, S.Z., Khan, M., Alam, A.N.: Production of chitin and chitosan from shrimp shell wastes. J. Bangladesh Agric. Univ. **14**(2), 253–259 (2016)
42. Lacoste, C., Lopez-Cuesta, J.M., Bergeret, A.: Development of a biobased superabsorbent polymer from recycled cellulose for diapers applications. Eur. Polymer J. **116**, 38–44 (2019)
43. Roy, J.C., Salaün, F., Giraud, S., Ferri, A., Chen, G., Guan, J.: Solubility of chitin: solvents, solution behaviors and their related mechanisms. Solubility Polysaccharides **3**, 20–60 (2017)
44. Pillai, C.K.S., Paul, W., Sharma, C.P.: Chitin and chitosan polymers: chemistry, solubility and fiber formation. Prog. Polym. Sci. **34**, 641–678 (2009)
45. Mir, M., Ali, M.N., Barakullah, A., Gulzar, A., Arshad, M., Fatima, S., Asad, M.: Synthetic polymeric biomaterials for wound healing: a review. Prog. Biomater. **7**(1), 1–21 (2018)
46. Naseri-Nosar, M., Ziora, Z.M.: Wound dressings from naturally-occurring polymers: a review on homopolysaccharide-based composites. Carbohyd. Polym. **189**, 379–398 (2018)
47. Sagar, B., Hamlyn, P., Wales, D. (1994). Wound dressing. European Patent 0291587

Alginate-Based Superabsorbents

D. Thirumoolan, T. Siva, R. Ananthakumar, and K. S. Nathiga Nambi

Abstract Superabsorbent polymers (SAPs) are widely and increasingly used in various applications. These kinds of naturally occurring biopolymers have more attention as a function of efficient absorbents. This is due to their biocompatibility and biodegradability, low-cost production, and ease of functionalization. Among them, alginate-based absorbents are considered one of the most effective absorbents. The presence of active acid and hydroxyl groups along the alginate backbone facilitates its physical and chemical modifications and participates in various possible absorption mechanisms. Herein, we focus our attention on presenting the recent advances in alginate-based superabsorbents, including preparation, fragmentation, and applications. Also, the modification of alginate by various materials and composites is addressed. The book chapter provides essential information to understand the theories behind the superabsorption mechanisms, which are clearly discussed. Finally, the chapter addresses the future challenges and other potential applications of alginate-based superabsorbents.

D. Thirumoolan (✉)
Department of Chemistry, Annai College of Arts and Science (Affiliated to Bharathidasan University), Kovilacheri, Kumbakonam 612503, Tamil Nadu, India
e-mail: thirumooland@gmail.com

T. Siva · R. Ananthakumar
Laboratory for Advanced Research in Polymeric Materials (LARPM), School for Advanced Research in Petrochemicals, Central Institute of Petrochemicals Engineering and Technology (CIPET), Bhubaneswar 751024, India

Advanced Research School for Technology and Product Simulation (ARSTPS), School for Advanced Research in Polymers (SARP), Central Institute of Petrochemical Engineering & Technology (CIPET), Tamil nadu 600032 Chennai, India

T. Siva
e-mail: chemistry_siva@yahoo.co.in

R. Ananthakumar
e-mail: ananth@cipet.gov.in

K. S. N. Nambi
Department of Biology, The Gandhigram Rural Institute (Deemed to Be University), Gandhigram, Dindigul District 624302, Tamil Nadu, India
e-mail: ksnambi@gmail.com

S. Pradhan and S. Mohanty (eds.), *Bio-based Superabsorbents*, Engineering Materials,
https://doi.org/10.1007/978-981-99-3094-4_6

Keywords Superabsorbent · Biopolymer · Alginate · Fabrication · Application

1 Overview of Superabsorbent Polymers

A polymer that can absorb and hold a significant amount of liquid in comparison to its own mass. Superabsorbent Polymers (SAPs) were originally used in agriculture and diaper manufacturing over four decades ago [1]. While a superb water-absorbent property was of the utmost importance at the time, SAP has since expanded its applications to other industries. SAPs are cross-linked, hydrophilic polymer networks that absorb a substantial amount of water or aqueous saline solutions. In a relatively short length of time, SAP absorbs water or aqueous saline solutions roughly 10–1000 times its initial weight or volume [2]. Superabsorbent polymers, often known as SAP, are a type of intelligent and functional polymer that was first introduced to the science of polymers in 1938. These polymers are finding widespread and expanding use in a number of different applications. These types of natural biopolymers have more concentration as an effective absorbing function.

Recent research focuses attention on superabsorbent polymers based on natural polysaccharides for their unique biocompatible, biodegradable, renewable, and non-toxic properties. The polymer obtained in this way has very different characteristics than the individual materials. For example, the electrical conductivity of the resulting hydrogel has improved a great deal compared to that of the bare hydrogel [3]. A class of hydrophilic polymer materials known as "hydrogeological products" are able to retain a significant quantity of water in their 3D networks. Hydrogeological goods are what these materials fall under. The significant use of these goods in a variety of manufacturing and biological applications is universally considered to be of the highest importance. Because synthetic hydrogels have a greater ability to absorb water, a longer lifespan, and access to a wider range of chemical raw materials than natural hydrogels, they have steadily displaced their natural counterparts in the hydrogel market.

Among the hydrophilic polymers most often used in the preparation of hydrogels, there were numerous advantages of polysaccharides over synthetic polymers. Researchers are interested in polysaccharide-based hydrogels due to their potential in biomedicine and other sectors, such as pharmaceuticals, chemical engineering, agriculture, and food [3, 4].

Polymerization and progressive cross-linking are multistep processes. First, generate reactive polymer molecules, then cross-link them using the proper agents. Multiple step processes include polymerization and cross-linking of multifunctional monomers. Polymer engineers may design and manufacture structures with molecular-level qualities including biodegradability, mechanical strength, and chemical and biological reactivity [4, 5].

Among them, alginate-based absorbers are considered to be one of the most efficient. The existence of active acid and hydroxyl groups somewhere along the alginate spine makes it easier for the molecule to undergo physical and chemical changes.

These groups also take part in a range of different absorption methods that may be feasible. In this context, we focus on presenting recent advancements in alginate-based superconductors, including preparation, fragmentation, and applications. In addition, the modification of alginate by a variety of materials and composites is discussed. The chapter of the book provides vital information for understanding the theories underlying the mechanisms of superabsorption, which are clearly discussed. Finally, the chapter looks at the challenges ahead and other potential applications of alginate-based superabsorbers.

1.1 Classification of Superabsorbent Polymers

The Superabsorbent Polymers may be classified according to multiple roles/functions as discussed in upcoming sections. Figure 1 shows the classification of SAPs.

1.1.1 Based on Origin

The source denotes a single entity or individual that receives something. Consequently, depending on their origin, these polymers may fall into either the synthetic or natural category. Biocompatibility and biodegradability of natural polymers are cost-effective, but their mechanical qualities are markedly poorer. However, there is a deficiency of bioactive characteristics in synthetic polymeric materials [6].

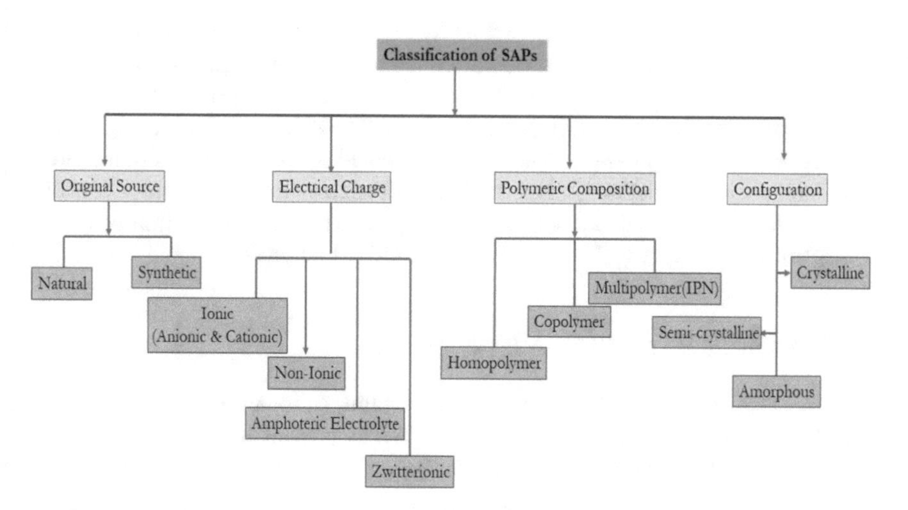

Fig. 1 Flow chart for the classification of superabsorbent polymers

1.1.2 Based on Configuration

In accordance with its morphological properties, SAP can alternatively be categorized as (a) an amorphous substance (non-crystal) (no obvious organization), (b) Semi-crystalline: A material that is a complicated blend of crystalline and amorphous phases (characterized by improper crystallization), (c) crystalline (regular atomic arrangement in the space grid), and (d) fibers.

The copolymer hydrogel that was deduced from N,N-dimethyl acrylamide and n-octadecyl acrylate, for instance, was described as semicrystalline and evinced thaw, high durability, self-healing, and shape-memory properties. Self-healing and shape remembering capabilities were further important features. Employing poly(N-isopropyl acrylamide) and N-hydroxymethyl acrylamide, an autocross-linking crystal microgel was produced. The nanoparticles of silver were encapsulated in carboxymethyl cellulose that generate an amorphous hydrogel and used in the healing process.

1.1.3 Based on Polymer Composition

- The term "homopolymeric SAP" refers to polymer systems that start from a single monomer species that consist of a characteristic auxiliary fragment in any polymer framework. Cross-linked morphological structures may be predominantly focused on the polymerization method that is applied as well as the type of monomer that is used. Depending on their tacticity, linear superabsorbent homopolymers can be isostatic, syndiotactic, or atactic.
- Copolymeric SAPs consist of two or more distinct monomer species, each of which contains at least one hydrophilic component, and are organized in a random, block, or alternating manner throughout the polymer chain [7].
- Multipolymers Interpenetrating polymeric SAPs (IPN) are a significant subclass of superabsorbent polymers. They are constructed of two various artificial and/or natural polymers that are joined together and envelop one another in a network. The cross-linked and uncross-linked polymers that make up the Semi-IPN superabsorbent hydrogel are two separate kinds of polymers.

1.1.4 Based on the Electrical Charge

They are also able to be classified as either neutral, cationic, or anionic polymers according to the amount of electrical load that is present in the SAP along the polymer's backbone and/or side chains [6, 8].

- Polymers that have no net electrical charge are said to be neutral, or nonionic. Polymers that are examples of this group include hydroxyethyl cellulose that has been grafted onto polyacrylamide and SAP which is based on hydroxymethyl cellulose. Due to the existence of hydrogen bonds, water molecules may dissolve in both of these polymers. In addition, a rise in the entropy of the polymer–solvent system during the blending process would enhance the polymer's ability to absorb water.

- Ionic: SAPs with anionic or cationic moiety. The potassium, ammonium, and sodium salts of acrylic monomers will be anionic. Cationic SAP will be produced from chitosan and acrylic monomers.
- Zwitterionic (polybetain) SAPs that have both anionic and cationic groups but have no overall charge are said to as neutral. In the context of "intelligent" electrosensitive SAPs, this categorization is applied.
- Ampholytics: These SAPs contain both acidic and core characteristics as structural components.

1.2 General Swelling Mechanism of SAPs

The SAPs are compounds capable of absorbing and retaining vast volumes of water and aqueous solutions. Consequently, they are ideal for applications needing water absorption. Previously, SAPs were made up of hydrophilic polymers with a high affinity for water, such as modified starch, cellulose, poly(vinyl alcohol), and poly(ethylene glycol) (ethylene oxide). After being cross-linked chemically or physically, these polymers become water swellable but not water-soluble. Modern SAPs are made from partly neutralized, slightly cross-linked poly (acrylic acid), which offers the greatest performance-to-cost ratio. The polymers are produced at low solids levels for both quality and economic reasons and are subsequently dried and processed into white granular solids. In general, SAPs are networks of flexible polymer chains that have been cross-linked. Prior to water absorption, the polymer chains are contracted. Modern SAPs are composed of partially neutralized, somewhat cross-linked poly (acrylic acid), which provides the best performance to cost ratio. Within the network, water is firmly confined. In water, they turn into a flexible gel that is up to 99% in some cases. The process of swelling and the final swelling capacity is affected by a number of factors, such as the network structure, the type of fluid, the structure of the hydrogel (porous or not porous), and the way it is dried.

In water, oxygen is electronegative and attracts hydrogen's electrons, creating a dipole. Hydrogen bonds form when hydrogen atoms connect with tiny electronegative atoms like N, F, or O. Hydrogen atoms bind to nearby nonbonding electron pairs [3]. Water's electronegative atom, oxygen, attracts hydrogen's electrons to form a dipole. Hydrogen atoms are attached to water molecules' oxygen lone pairs. Oxygen contains two unpaired electron pairs that may form hydrogen bonds [9]. This reduces resources and increases system entropy. Due to SAP's hydrophilic nature, polymer chains spread in water, increasing configurations and entropy [10]. Water-insoluble SAPs are provided by cross-links between polymer chains in a 3D network, which inhibits degradation and expansion. In addition, when the network's elastic retraction forces tighten the links, their entropy decreases. There is now an equilibrium between the forces of contraction and the chains' inclination to extend forever. Increased cross-link density is inversely proportional to gel swelling capacity and directly proportional to gel strength.

The interaction between solvent and polymer in the case of ionic polymers extends beyond simple mixing. The neutralized chains have contrary charges (Fig. 2). Positive sodium ions balance negative carboxylate groups, preserving electrical neutrality. Water hydrates sodium ions. Their attraction to carboxylate ions decreases (due to the high dielectric constant of water). This encourages the free movement of sodium ions, which boosts the gel's osmotic pressure. However, mobility of sodium ions are unable to exit the gel due to their weak attraction to the negative carboxylate ions present on the polymer backbone. Consequently, sodium ions act as if they were trapped behind a semipermeable barrier. Consequently, the driving force for swelling is the pressure difference between the interior and exterior of the gel. The gel's osmotic pressure and swelling capacity will decrease as the concentration of salt outside the gel increases. The gel will grow to its greatest degree in deionized water [11, 12].

2 Overview of Alginates-Based Superabsorbent

Alginate is a naturally occurring anionic polymer that is often found in the cell walls of brown algae (Phaeophyceae). Alginate is the primary structural component of seaweed. Similar to cellulose in terrestrial plants, the gel in the cell walls and intercellular matrix provides mechanical strength and flexibility to seaweed [13]. In addition, the varied alginate compositions of different algae reflect this activity. 20–60% of dry matter is alginate, while the brown algae have an average of 40% alginate. Brown seaweed comprises alginate gels containing barium, calcium, magnesium, sodium, and strontium ions [14]. All commercial alginate is generally obtained from algal biomass [15]. Alginate isn't only brown algae.

2.1 Structure of Alginates

The biopolymer family alginate is unbranched. Partial acid hydrolysis reveals that alginates are composed of 1,4-L-guluronic acid (G) and 1,4-D-mannuronic acid (M), either heterogeneous block (MG) or homogeneous block (HB) (poly-M, poly-G). Each alginate-producing species may have distinct alginate compositions and varying ratios of mannuronic to guluronic acid blocks [16]. These two acids differ in seaweed species and location [17].

Alginates do not contain regular repeating units, and the distribution of monomers in the polymer chain cannot be described. Insufficient understanding of monomeric composition prevented the prediction of the sequential structure of alginates from various species.

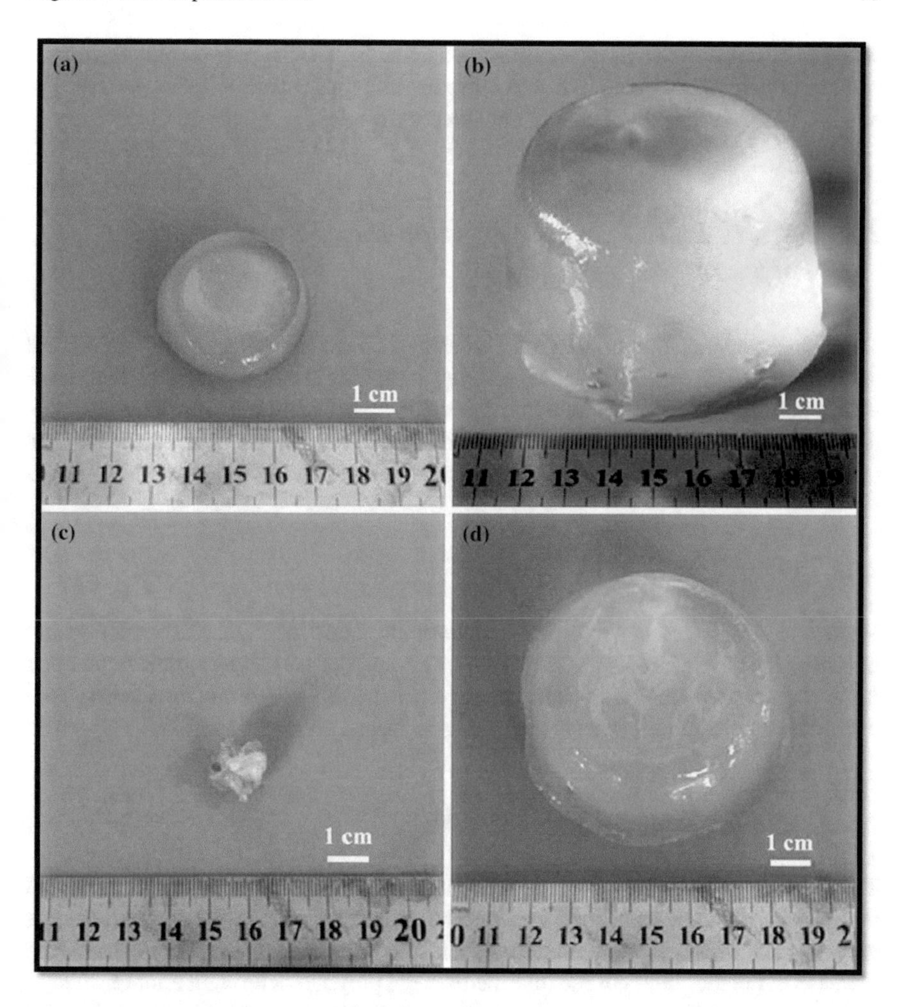

Fig. 2 Swelling mechanism of SAPs [10]. **a** Original hydrogel, **b** swollen hydrogel, **c** dried hydrogel, and **d** hydrogel after swelling in NaCl solution for a week. Copyright from [10]

Alginates are extracted from the cell walls of three brown algae species *Macrocystis pyrifera*, *Laminaria hyperborean*, *Ascophyllum nodosum*, and numerous bacteria (*Azotobacter vinelandii*, *Pseudomonas spp.*) [13]. Seaweeds can benefit from the structural and ion-exchange capabilities of this compound. Polymannuronic acid-rich alginates are detected in embryonic cell wall tissue and/or intercellular locations, whereas polyguluronic acid-rich alginates are found in cell walls with a high affinity for Ca^{2+} [13]. Alginates in cell walls have a high affinity for Ca^{2+}, the fundamental determinant in gel strength. The cytoplasm is responsible for the production of the alginate polymer, which is eventually exported from the cell [18].

This word refers to any derivatives of alginic acid as well as the salts of such derivatives. Producing alginic acid involves utilizing HCl that has been diluted together with algae. The filtrate extract is then treated with NaCl or $CaCl_2$ to precipitate fibrous sodium or calcium alginate. After treating the precipitate with an acid, purifying it, and lyophilizing it, sodium alginate powder is formed.

2.2 Factors Influencing the Alginate-Based Superabsorbent Polymers

To produce superabsorbent, free radicals catalyze the polymerization of acrylic acid and its metal salts in an aqueous solution including a cross-linker. In the aforementioned procedure, the monomer, its concentration, the initiator, its concentration, the cross-linker, its concentration, the monomers' respective reactivities, the polymerization kinetics, and the reaction temperature are all critical [19].

2.2.1 Effect of Materials

We would want to select a remarkable material for constructing the 3D structure and performance base of SAPs based on its exceptional features. Hydrophilic nature and water absorption capabilities, such as capillary and swelling, are crucial. Choose from cellulose, starch, polysaccharides, synthetic resin, and organic–inorganic composites.

2.2.2 Effect of Monomer Concentration

The features of the polymer that is produced, as well as the rate of the reaction and the amount of money that can be made from it, are all affected by the concentration of the monomer in the reaction mixture. As a result of the process of polymerization, an increase in the toughness of the intermediate gel polymer is brought about by a high monomer concentration. The gel's tensile strength influences equipment design, gel particle size during agitation and heat removal procedure. Chain transfer to polymer increases with increasing monomer concentration, especially at high conversion, resulting in branching and self cross-linking activities that change product properties. To counteract these adverse effects, chain transfer agents are beneficial. The effective use of the cross-linker is another factor that determines monomer concentration. Because the solubility of the cross-linking agent increases as the monomer concentration rises, the cross-linker becomes more effective. As monomer concentration increases, network cyclization decreases [20].

2.2.3 Effect of Initiators

With the help of thermally decomposable initiators, redox initiators, or any combination of the two, the polymerization process is kicked off with the production of free radicals in the aqueous phase. Thermal initiators include potassium persulfate, ammonium persulfate, and 2,2′-azobis (4-cyanopetanoic acid). Couples of persulfate/ascorbate, persulfate/bisulfate, and persulfate/thiosulfate are assumed to be the components that make up the redox system that was once thought to have been responsible for cross-linking copolymerizations.

2.2.4 Effect of Neutralizing Agents

Monomers and cross-linkers are dissolved in 10–70% water. Active substances for neutralization include sodium hydroxide and sodium carbonate. The pH of the base solution and the potential for hydrolyzing the cross-linker, the base's solubility in water, and the monomer salt's solubility in water must be considered. This allows for the finest decision. Due to acrylic acid's solubility in hydrocarbons, the monomer must be neutralized during suspension (the continuous phase). Branching and self cross-linking increase with monomer concentration, especially at high conversion. Chain-transfer agents can reduce these harms. Cross-linker efficiency also affects monomer concentration. Cross-linker efficiency increases as cross-linker solubility grows with monomer concentration. Network cyclization decreases with monomer concentration. Chain-transfer agents can reduce these harms. Cross-linker efficiency also affects monomer concentration. Cross-linker efficiency increases as cross-linker solubility grows with monomer concentration. Network cyclization decreases with monomer concentration.

2.2.5 Effect of Cross-linking Agents

Cross-linkers, even in trace levels, have an important role in modifying the properties of superabsorbent polymers. 1,4-butane diacrylate, triallyl cyanurate, 1,2-ethylene glycol dimethacrylate, diallyl phthalate, 1,4-butanediol diacrylate, 1,6-hexanediol diacrylate, N,N-methylene-bis-acrylamide, 1,2-ethylene glycol diacrylate, triallyl amine, and other di- and tri-acrylate esters and amides are extensively used. Cross-linker impacts soluble polymer formation, swelling, and mechanical properties.

The proportion of cross-linker to acrylic acid or sodium acrylate will affect the propensity of the cross-initiating linker to deplete during polymerization. Early cross-linker depletion ought to boost the solubility of the final product.

Neutralizing carboxylic acid groups affects cross-linker choice. In solution polymerization, cross-linker solubility impacts cross-linking efficiency. Steric hindrance influences cross-linking and mobility at a pendent double bond.

2.2.6 Effect of Temperature

When a system's temperature fluctuates, superabsorbent polymers are used. In a diaper, superabsorbent polymer is soaked in a salt and urea solution at body temperature, but the resulting gel rapidly cools outside. Climate and other environmental variables will affect the rate of cooling. Since the polymer solution diffusion coefficient fluctuates with temperature, superabsorbent polymers should reflect this.

2.2.7 Effect of Physical and Chemical Factors

Sodium alginate is soluble in hydrochloric and sulphuric acid but not ethanol, methanol, DMSO, acetone, chloroform, dichloromethane, or n-butanol. Generally, sodium alginate is slowly soluble in water and generates a viscous solution. Sodium alginate is soluble in cold and hot water but insoluble in diethyl ether.

Also, it affect the viscosity by pH factors. A reduction in pH enhances viscosity by protonating carboxylate groups in the alginate spine, which creates hydrogen bonds. The typical amount of molecules in commercial alginate is 33,000–400,000 g/mol. Increasing alginate's molecular weight changes gel properties (for instance, viscosity). Unlike ALG monovalent salts and alginate esters, alginic acid is insoluble in water and organic solvents. Alginate with poly-M or poly-G structures precipitates at low pH, although others are soluble.

Alginates can be functionalized at C-2, C-3, or –COOH (C-6). Using the different reactivities of the two functional groups, alginic acid may be modified at either location. Due to minute differences in reactivity, it's difficult to selectively modify the C-2 or C-3 hydroxyl group.

The distribution of M, G, or MG blocks and, indirectly, the M/G ratio are both important factors that influence the pattern of substitution or functionalization that occurs in alginates. It is challenging to characterize the structural makeup of derivatives when there are no alginates with regulated sequences that are readily accessible for purchase in the marketplace.

3 Synthesis of Alginate-Based Superabsorbent Polymers

The vast majority of SAPs are either entirely synthetic or derived from petrochemicals. SAPs are produced using several monomers, typically acrylics. In the majority of cases, acrylic acid (AA) and its sodium or potassium salts, as well as acrylamide (AM), are used to produce SAPs. Due to their better cost-to-performance ratio [21, 22], the bulk of superabsorbents is presently fabricated from synthetic polymers (mostly acrylics). However, the worldwide firm decision for environmental preservation may promote the notion of replacing synthetics partially or entirely with "greener" substitutes [23]. Natural polymers are the organic resources that are least costly, most environmentally friendly, and most abundant (polysaccharides). Alginate, starch, chitosan, cellulose, pectin, and gelatin are all examples of hydrophilic prepolymers that may be found in naturally occurring substances. Despite the fact that they are difficult to handle, compounds with glycosidic or amino acid repeating units are the subject of a significant amount of research and are used in a wide variety of applications. These polysaccharide-based SAPs are more sustainable, biodegradable, and non-toxic than their synthetic alternatives.

The design of SAPs is based on the monomer and Cross-linker types, which have a substantial effect on their characteristics and applications. Bulk, solution, and suspension polymerizations are the most prevalent SAP manufacturing processes. In addition, polymerization by radiation has gained substantial interest for future clean technology production. Cross-linking polymerization has a substantial influence on the properties of superabsorbent polymers. Describe the various alginate-based superabsorbent synthesis techniques in this section.

3.1 Extraction from Natural Source

Alginate extraction requires pre-extraction, neutralization, and precipitation [24–26] (Fig. 3). Seaweed was properly cleaned and air-dried in the shade. Dry seaweed was coarsely crushed and sieved to remove seaweed fiber. The milled seaweed was weighed and immersed for 24 h in 2% formaldehyde. The formaldehyde solution was filtered and rinsed with distilled water after 24 h. In the pre-extraction phase, Na(I), Ca(II), Mg(II), Sr(II), and Ba(II) counter ions are eliminated. Neutralization of insoluble alginic acid with sodium carbonate or sodium hydroxide produces soluble sodium alginate, which is subsequently separated by sifting, flotation, centrifugation, and filtration. Adding alcohol, calcium chloride, or a mineral acid precipitates sodium alginate. Dry milling the sodium alginate precipitates.

3.2 Polymerisation Method

Polymerization, cross-linking, and co-grafting are all advantageous ways for improving its mechanical and functional properties. To improve the physical and mechanical properties of synthetic components, natural polymers are blended with synthetic polymers. Through chemical modification, derivatization, or co-grafting, alginates' solubility, hydrophobicity, physicochemical, and biological characteristics can be changed. A graft copolymer has a linear backbone of one monomer with one or more branches (grafts) of another monomer distributed at random. Both the main chain and the graft chains may be homopolymeric or copolymeric.

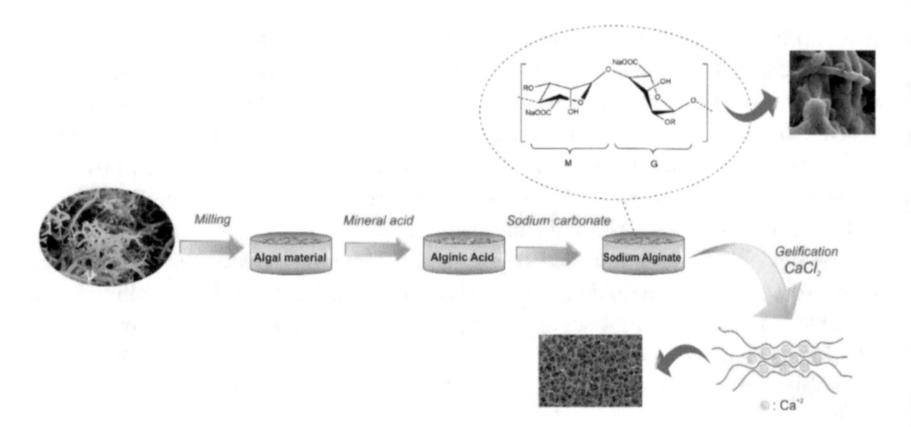

Fig. 3 A typical process for the extraction of sodium alginate from brown algae [25]. Copyright from [25]

In this procedure, NaAlg was combined with distilled water in a flask with three necks [27]. The reactor was put in a thermostatically controlled water bath at the desired temperature (70–90 °C) for 20 min. After dispersing NaAlg and homogenizing the mixture, a tiny quantity of an initiator (Ammonium Persulfate or Potassium persulfate) is added and the mixture is agitated for 15 min. Gelation is seen 10–15 min after monomers (Acrylic acid, Acrylamide, or) and cross-linker solutions are applied concurrently. At the completion of the propagation process, the gel product was put in ethanol and dehydrated for 48 h. The product was then cut into little bits, repeatedly washed with ethanol, and filtered. In an oven, the finished product was dried for 24 h.

3.3 Photo Chemical Method

Biomedical applications encourage the usage of photo cross-linked substances. In a photochemical reactor, photo graft-copolymerization processes took place. With minimally invasive methods, cells and/or bioactive components can be added to aqueous macromer solutions that can then be cross-linked in place after a short exposure to ultraviolet (UV) or visible light. Also, photoinitiators like ceric ammonium nitrate, which make free radicals, can be added to speed up the polymerization process and turn the macromer solution into hydrogels.

Sodium salt of partially carboxylated sodium alginate (Na-PCMSA) (0.2–3.0 g, dry basis) was dissolved in low conductivity water (so the total volume of the reaction system stayed at 150 mL) and stirred at 35 °C for 1 h, then at room temperature for 20 min. Freshly made CAN solution (0.5×10^{-3} to 10.00×10^{-3} mol/L) in 10 mL nitric acid was added to the reaction flask, which was then flushed with filtered nitrogen gas for 30 min. In the photochemical reactor, the reaction flask was exposed to a 125 W medium pressure mercury lamp while a steady flow of nitrogen gas and stirring went on for 0.5–10 h at 15–45 °C. The rough product was spun in a centrifuge, washed with diluted nitric acid and 95% methanol, and then washed with pure methanol. At 40 °C, the Na-PCMSA-g-PAN copolymer dried [28].

3.4 Radiation Method

In order to initiate the polymerization of unsaturated compounds into SAPs, high-energy ionizing radiation is used. This radiation can take the form of gamma rays or electron beams. The formation of radicals on the polymer chains is caused by the exposure of the polymer solution in water to radiation. Hydroxyl radicals are produced when water molecules are radiolyzed; these radicals subsequently attack polymer chains, which results in the generation of macro radicals. The outcome of the recombination of these macro radicals into covalent bonds is the formation of a structure that is cross-linked. The product that is acquired using this method is relatively pure in comparison to the product that is obtained using the chemical initiation approach. This is due to the fact that there is no chemical initiator involved in this method.

Prior to irradiation, aqueous solutions had the oxygen removed by vacuum, and then the solutions were placed in glass vials and sealed. This was done to prevent the alginate from degrading. Samples were irradiated at the Atomic Centre of Ezeiza (CNEA, Argentina) with ^{60}Co gamma rays [29]. Lyophilization followed by a Soxhlet extraction in methanol was used to remove any unreacted N-isopropyl acrylamide (NIPAAm) monomer and free poly-N-isopropyl acrylamide (PNIPAAm) homopolymer from the samples after irradiation. Following this procedure, we made certain that the PNIPAAm that was still present in the final product corresponded to molecules that had been grafted onto the alginate backbone. The samples were then dried until they reached a consistent weight.

4 Applications of Alginate-Based Superabsorbents

The many fields and businesses that might benefit from alginate-based superabsorbent are extensive. As shown in Fig. 4, alginate-based hydrogels can be used in a wide variety of contexts [30].

4.1 Antiviral Activity

Materials made from alginate have low or no toxicity and can stop a wide range of viruses. In humans, these viruses include hepatitis A, B, and C, herpes simplex

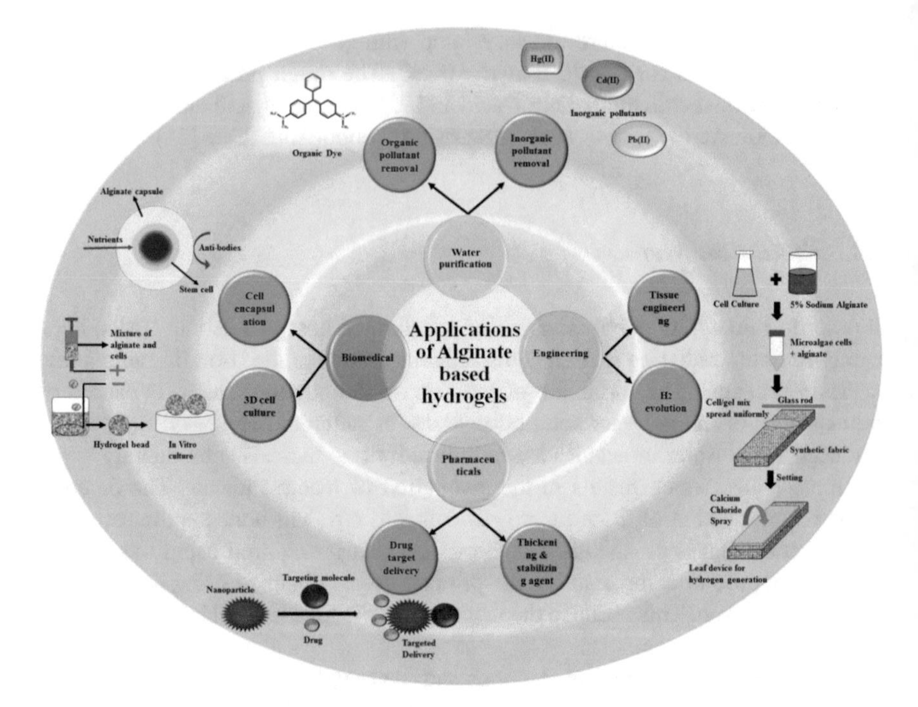

Fig. 4 Various applied applications of alginate based hydrogels [30]. Copyright from [30]

virus types 1 and 2, Sindbis virus, rabies virus, rubella virus, poliovirus type 1, influenza virus, and human immunodeficiency virus type 1 [31]. In rodents, these viruses include murine norovirus, and in plants, they include potato virus X and viruses with envelopes, Calcium alginate may kill viruses because it has negatively charged particles that stick to viral envelopes and stop membrane receptors from working [32].

4.2 Wound Dressing

Even though these materials are now part of the body, they have been used to replace tissues for more than 20 years. The Food and Drug Administration (FDA) has approved alginate for medical uses, and it has been used to treat wounds without any signs of harm. Superabsorbent polymers (SAPs)/hydrogels that are naturally porous and have a water content of 85% can be used to replace the skin. SAP wound coverings are good because they keep wounds moist, help blood clot better, and have a lot of potential for tissue regeneration [33]. Amorphous hydrogels are soft gels that absorb water and become less thick as they do so. Hydrogel sheets are stable sheets that expand as they absorb fluid, yet they preserve their structural integrity and offer a moist environment that promotes wound healing.

The broader use of SAPs in wound dressings [6] was due to:

(i) High absorptive capacity, remains intact after equilibrium swelling.
(ii) Ability to produce osmotic gradient.
(iii) Biocompatibility.
(iv) The capacity to grab harmful ions.
(v) Gel-forming capacity.
(vi) Capable of absorbing and locking fluids, bacteria, matrix metalloproteinases (MMPs), and other enzymes within the dressing.

4.3 Drug Delivery

SAP hydrogels are often used as carriers for a wide range of drugs that need to be released slowly. Alginate hydrogels have the potential to be used as drug carriers for the sustained or localized delivery of low molecular weight drugs and macromolecules, such as proteins and genes, because of their unique properties, including high water content, nontoxicity, soft consistency, biocompatibility, and biodegradability [34, 35]. The perforations in a carrier have the potential to entrap or disable its load. Which gel alginate forms is determined by the pH of the surrounding media. Contraction and production of an acidic gel at low pH (gastrointestinal environment) inhibit drug release. Alginic acid's skin-like structure transforms into a soluble viscous gel after passing through the digestive system with a higher pH, causing medicine breakdown and release. Systemic toxicity can be prevented by releasing medications slowly into a target tissue. Drug release from hydrogel pores is regulated by diffusion, swelling, chemistry, and the environment.

Creating, modifying, and characterizing alginate hydrogels for in-vivo use. Under different settings, the drug release properties of hydrogel beads containing diclofenac sodium (DS) were examined. At pH 2.1, diclofenac sodium (DS) release was less than 5% after 24 h, whereas at pH 6.8 it was 100%. In a medium with a pH of 7.4, nearly all of the DS was released in about two hours. Since composite hydrogel beads disintegrate at a lower pH (pH 2.1), DS migration is challenging. Expanding composite hydrogel beads provide more room for DS migration, which results in enhanced drug release at higher pH levels (pH 6.8 or 7.4) [36].

In the realm of pharmacology, alginate hydrogels are employed for a variety of purposes, such as emulsion stabilization, suspending agents, tablet binders, and tablet disintegrants [37]. It is conceivable to investigate the use of microcapsules composed of alginate composites to transport medications. They are typically created by layering successively negatively and positively charged polyelectrolytes. For instance, a microcapsule that is biocompatible and can contain positively charged molecules is created by sequentially breaking down alginate and chitosan on $CaCO_3$ particles, followed by the removal of the core.

4.4 3D Bioprinting

Bioink is a term used to describe biomaterials that may contain cells and are capable of being printed as 3D scaffolds and tissue-like structures. It is feasible to duplicate or replace the target tissue based on its similarity to the original cell membrane and physical properties, as well as its customizable degradability.

The biocompatible polyanionic and linear block copolymer sodium alginate encourages cell development. Because of its shear-thinning properties, alginate solution is a great starting material for 3D bioprinted tissue engineered structures [38, 39]. A low-viscosity alginate solution was developed to get around the problems associated with in situ cross-linking. As a result, alginate hydrogels could be extruded using shear from a syringe and would quickly reconstitute if the mechanical force was removed. It will be challenging to ensure the structural and shape fidelity of the resulting hydrogels if alginate solutions are employed as the bioink to generate the scaffolds. This is because the viscosity of the alginate solutions is reduced at the maximum concentration, making it more likely that the filaments would collapse and cling together. Cross-linking alginate hydrogels with Ca^{2+} makes them weaker and more likely to break under the weight of gravity. This changes their structure and throws off the accuracy of hydrogels that are printed in three dimensions.

4.5 Cosmetics

Alginates are a cosmetic important component. In the field of cosmetics, they are utilized as emulsifiers, consistency enhancers, and thickening agents in order to provide a top layer that is capable of retaining moisture. They protect the skin by hydrating it since they can retain water. Insoluble alginates expand in water. It can take up water hundreds of times quicker than hyaluronic acid. These ingredients can be found in a wide variety of personal care products, including toothpaste, hand lotions, ointment bases, pomades, and greaseless creams. Aging and skin problems are both thwarted by alginate's presence. Keeping the organoleptic properties of cosmetics is the job of antioxidants, which do so by avoiding lipid oxidation [40–42]. This reduces the ability to see or smell anything different.

4.6 Agriculture

A standard way of applying agrochemicals results in an initial concentration that is significantly higher than the levels of active components. Because of this, conventional methods of applying agrochemicals offer a higher initial concentration than is strictly necessary. Because of their high level of solubility in water and their ability to diffuse into the environment around them, crops are unable to make use of approximately ninety percent of the agrochemicals that are distributed using conventional methods [43, 44]. The sustained release is a method or approach in which active chemicals or water are made available to a specific target at a set pace and time interval. This is done so that the effect that was intended happens. SAPs are things that can soak up 100–1,000 times their own weight in water. In addition to being able to soak up water, SAPs are used for the slow release of nutrients. When the ground starts to dry up, SAP gives the water it gathered during the wet season to the soil around its roots. This means that the soil dries out after the SAP does. Soil that has SAP added to it keeps water for a week longer than soil that doesn't have SAP.

5 Special Features of Alginate-Based Superabsorbents

Alginate is beneficial in food, biomedicine, electronics, and environmental remediation due to its biocompatibility and mild gelation conditions. In environmental treatment, alginates are utilized. Currently, superabsorbent polymer (SAP) manufacturers are unable to fulfill worldwide demand due to the increasing number of elderly and infants. In addition, manufacturers must expand their SAP software manufacturing capacity to fulfill present and future demands. Enhance SAP's usefulness in a variety of applications and increase its more recent, long standing attributes (such as high gel strength, super absorbing capacity, and faster absorption rate).

Due to the lack of information on SAP's environmental degradation and its impacts, it was unclear whether it posed a threat to the environment or soil. Additionally, the rate (dosage) and manner of SAP use in agriculture must be investigated. Determine the benefit-to-cost ratio of several types of agricultural SAP. Long-term field research must be conducted on SAPs, preferably in conjunction with changes in soil mechanical characteristics. The experts disregard the amount of soil and the remaining monomer content of SAP. This issue plagues SAP programs that operate for an extended period. The researchers ceased their investigation into the capacity of freshly synthesized SAP materials to absorb water under load and salt. SAPs must be reformulated after several cycles of absorption and desorption to preserve their mechanical strength. This will enhance cycle results. High-performance and ultrahigh-strength SAPs and superabsorbent polymer composite (SAPC) with high swellability, rapid absorption, enhanced gel strength, and reduced environmental impact may be manufactured with the use of nanotechnology because of its role in the fabrication of superior materials [45]. Contributions made by nanotechnology to the manufacturing of improved materials may play a role in the production of high-performance and ultrahigh-performance materials. Polymer fillers and nanoparticle control are efficient means of achieving this goal.

In several applications, stimuli-responsive alginate-based polymers have gained popularity. The choice of metal ions influences the hydrogel material's stability, elasticity, and swelling when alginate polymers undergo ionic cross-linking. Metal ion selection influences these characteristics. Stimuli can influence the rate of alginate hydrogenation and the uniformity of hydrogel structure. Laboratory work is mandatory. Using stimuli like pH, light, electric field, and enzymes, metal ions from an inert source may be released to cross-link alginate chains. CO_2, DMSO, and carboxylic acid molecules can affect the cross-linking of alginate chains during hydrogel production, hence altering the characteristics of alginate. By functionalizing the alginate molecule, stimuli-responsive materials can be created. The chemical modification of the alginate chain can confer new properties on alginate-based products. By covalently attaching functional substituents to the alginate backbone, intelligent alginate materials for biological applications such as controlling medicine distribution, bioprinting, and tissue engineering may be manufactured. Fluorophores and biochemical anchors are functional substituents.

Target-specific applications, such as stabilizing labile molecules against degradation, site-specific drug targeting, reducing dosing frequency, and prolonging therapeutic outcomes, can benefit from the grafting of synthetic polymer chains to the alginate chain, alginate-based interpolymer networks (IPNs), or alginate-based nanomaterials. Grafting is possible when alginates are mixed with synthetic polymers. Alginate-based smart materials can be broken apart and their contained contents released in response to external stimuli. Alginate's functional groups are sensitive to stimuli, allowing for controlled degradation. Biosensors, self-healing materials, drug delivery, and tissue engineering have all made use of alginate materials that respond to a variety of stimuli, such as pH, biomolecules, light, enzymes, electrical impulses, and mechanical energy. The use of these materials in tissue engineering and drug delivery shows promise. Water purification and electrochemical cells for energy harvesting and storage are only two examples of how an alginate-based network might be used for environmental remediation by encapsulating (bio)molecules and electronic components [46]. There is some speculation that this system may be utilized to purify water.

Making intelligent alginate materials, which are crucial for meeting a broad variety of human demands, still faces challenges. Success in the future will depend on the thoughtful selection of reactive functional groups, their proportions, functional group kinds, and structural processes in the construction of alginate material. They are crucial to the future's success. It is necessary to adjust the physical, chemical, and biological properties of these substances to meet the needs of the recipient. Degradation is a key worry during functionalization, despite the fact that chemical modification of the alginate chain may be exploited to impart new capabilities or enhance innate characteristics in intelligent materials. By lowering alginate's weight, potent chemicals, acid or basic treatment, or both, might accelerate the backbone's disintegration [47]. To create functional alginate derivatives, chemicals and reaction conditions must be selected with care. Cross-linking and alginate chain modification strategies that can be used to create new materials with predictable reactions to a variety of stimuli require more study. Alginate materials paired with advanced delivery technologies may provide biological treatments that are safer and more effective than traditional medications. Cells can be harmed by alginate polymers and chemical encapsulation. Eliminating unreacted chemical reagents and unwanted by-products, as well as choosing cell-compatible chemicals, are essential for these applications. In addition, the impact of alginate-based products on human physiology must be researched.

Research into the mechanisms at play, process optimization, and the development of novel materials with novel properties are all necessary for the environmental uses of alginate-based materials in the future. There can be more field testing of these compounds because of the complexity of the solutions and/or effluents. The development of an alginate-based technology for cleaning up contaminated environments will require more study into the effectiveness of alginate-based materials in immobilization technology, particularly in the areas of microbial survival and in situ bioremediation of contaminated soil or groundwater. Research into (multi)responsive materials is expected to increase [6, 48], therefore, the potential function of intelligent alginate systems is exciting. Additional restraints in SAP's industry make market share expansion difficult. Variable costs and stringent regulations govern healthcare. To compare the absorption effectiveness of the various SAP materials generated, it is also important to use international standard experimental procedures for testing the absorption capacity of synthesized SAPs. This permits the absorption efficiency of SAP materials to be compared.

6 Conclusions

In this chapter, we looked at the impact on how well SAP functions in various aspects of its surroundings. Also, the structures of alginates and the components that influence superabsorbent polymers based on alginates are discussed. Further, the extraction of SAPs from natural sources and the synthesis of SAPs based on alginate utilising the techniques (polymerization process, photochemical method, and radiation method) are described. The real goal of this chapter is to improve academics' awareness of the many distinct SAP preparation tactics and the ways in which those strategies may be utilised in real-world applications. In addition to this, the innovative preparation of SAPs and SAP composites is useful for environmental-based long-lasting applications.

Acknowledgments The author, Dr. D. Thirumoolan, would like to express his gratitude to the management of Annai College of Arts and Science, Kovilacheri, Kumbakonam, Tamil Nadu, India, for providing him with the motivation to complete this project. Their support is greatly appreciated. One of the authors (T. Siva) thank Science & Engineering Research Board (SERB), New Delhi for financial support through the SERB National Post-Doctoral fellowship (2020–2021); F. No. PDF/2020/00181.

References

1. Omidian, H., Rocca, J.G., Park, K.: Advances in superporous hydrogels. J. Control. Release **102**, 3–12 (2005)
2. Ramazani-Harandi, M.J., Zohuriaan-Mehr, M.J., Yousefi, A.A., Ershad-Langroudi, A., Kabiri, K.: Rheological determination of the swollen gel strength of superabsorbent polymer hydrogels. Polym. Test. **25**, 470–474 (2006)
3. Wang, W., Wang, A.: Synthesis and swelling properties of pH-sensitive semi-IPN superabsorbent hydrogels based on sodium alginate-g-poly(sodium acrylate) and polyvinylpyrrolidone. Carbohyd. Polym. **80**, 1028–1036 (2010)
4. Wong, M.: Alginates in tissue engineering. In: Hollander, A.P., Hatton, P.V. (eds.) Biopolymer Methods in Tissue Engineering, pp. 77–86. Humana Press, Totowa, NJ (2004)
5. Hua, S., Wang, A.: Synthesis, characterization and swelling behaviors of sodium alginate-g-poly(acrylic acid)/sodium humate superabsorbent. Carbohyd. Polym. **75**, 79–84 (2009)
6. Venkatachalam, D., Kaliappa, S.: Superabsorbent polymers: a state-of-art review on their classification, synthesis, physicochemical properties, and applications. Rev. Chem. Eng., **39**(1),127-171 (2023)
7. Zohurian-Mehr, M.J., Kabiri, K.: Superabsorbent polymer materials: a review. Iranian Polym. J. **17**, 451–477 (2008)
8. Ahmed, E.M.: Hydrogel: preparation, characterization, and applications: a review. J. Adv. Res. **6**, 105–121 (2015)
9. Boyaci, T., Orakdogen, N.: Tuning the synthetic routes of dimethylaminoethyl methacrylate-based superabsorbent copolymer hydrogels containing sulfonate groups: elasticity, dynamic, and equilibrium swelling properties. Adv. Polym. Technol. **36**, 442–454 (2017)
10. Tang, H., Chen, H., Duan, B., Lu, A., Zhang, L.: Swelling behaviors of superabsorbent chitin/carboxymethylcellulose hydrogels. J. Mater. Sci. **49**, 2235–2242 (2014)
11. Zhang, H., Chong, Y.B., Zhao, Y., Buryak, A., Duan, F.: Self-sealing polyolefin by superabsorbent polymer. Eng. Sci. **8**, 66–75 (2019)
12. Pourjavadi, A., Barzegar, S.: Synthesis and evaluation of pH and thermosensitive pectin-based superabsorbent hydrogel for oral drug delivery systems. Starch Stärke **61**, 161–172 (2009)
13. Industrial Applications of Marine Biopolymers. CRC Press, Boca Raton (2017)
14. Haug, A., Smidsrød, O.: Strontium-calcium selectivity of alginates, Nature **215**(5102), 757 (1967)
15. Draget, K.I., Smidsrød, O., Skjåk-Bræk, G.: Alginates from Algae. Biopolymers online. In: A. Steinbüchel (ed.). Biopolymers Online, (2005). https://doi.org/10.1002/3527600035.bpol6008
16. Szekalska, M., Puciłowska, A., Szymańska, E., Ciosek, P., Winnicka, K.: Alginate: current use and future perspectives in pharmaceutical and biomedical applications. Int. J. Polym. Sci. **2016**, 7697031 (2016)
17. Haug, A., Larsen, B., Smidsrød, O.: Uronic acid sequence in alginate from different sources. Carbohyd. Res. **32**, 217–225 (1974)
18. Abe, K., Sakamoto, T., Sasaki, S.F., Nisizawa, K.: In vivo studies of the synthesis of alginic acid in Ishige okamurai. Bot. Mar. **16**, 229–234 (1973)
19. Zhang, W., Wang, P., Liu, S., Chen, J., Chen, R., He, X., Ma, G., Lei, Z.: Factors affecting the properties of superabsorbent polymer hydrogels and methods to improve their performance: a review. J. Mater. Sci. **56**, 16223–16242 (2021)
20. Santos, R.V.A., Costa, G.M.N., Pontes, K.V.: Development of tailor-made superabsorbent polymers: review of key aspects from raw material to kinetic model. J. Polym. Environ. **27**, 1861–1877 (2019)
21. Fredric, L.B., Andrew T.G., (eds.): Modern Superabsorbent Polymer Technology, John Wiley & Sons, (1997)
22. Pó, R.: Water-absorbent polymers: a patent survey. J. Macromol. Sci. Part C **34**, 607–662 (1994)
23. Salamone, J.C: Polymeric Materials Encyclopedia, Twelve Volume Set (1st ed.). CRC Press, (1996)

24. Torres, M.R., Sousa, A.P., Silva Filho, E.A., Melo, D.F., Feitosa, J.P., de Paula, R.C., Lima, M.G.: Extraction and physicochemical characterization of Sargassum vulgare alginate from Brazil. Carbohydr. Res. **342,** 2067–2074 (2007)
25. Abasalizadeh, F., Moghaddam, S.V., Alizadeh, E., Akbari, E., Kashani, E., Bagher Fazljou, S.M., Torbat, M., Akbarzadeh, A.: Alginate-based hydrogels as drug delivery vehicles in cancer treatment and their applications in wound dressing and 3D bioprinting. J. Biol. Eng. **14**, 1–22 (2020)
26. Saji, S., Hebden, A., Goswami, P., Du, C.: A brief review on the development of alginate extraction process and its sustainability. Sustainability. **14**(9), 5181 (2022)
27. Pourjavadi, A., Zeidabadi, F., Barzegar, S.: Alginate-based biodegradable superabsorbents as candidates for diclofenac sodium delivery systems. J. Appl. Polym. Sci. **118**, 2015–2023 (2010)
28. Trivedi, J., Chourasia, A., Trivedi, H.: Photo-induced synthesis, characterization and alkaline hydrolysis of sodium salt of partially carboxymethylated sodium alginate–g–poly(acrylonitrile). Res. Sq. 1–10 (2022)
29. Lencina, M.M.S., Rizzo, C., Demitri, C., Andreucetti, N., Maffezzoli, A.: Rheological analysis of thermo-responsive alginate/PNIPAAm graft copolymers synthesized by gamma radiation. Radiat. Phys. Chem. **156**, 38–43 (2019)
30. Thakur, S., Sharma, B., Verma, A., Chaudhary, J., Tamulevicius, S., Thakur, V.: Recent progress in sodium alginate based sustainable hydrogels for environmental applications. J. Clean. Prod. **198**, 143–159 (2018)
31. Serrano-Aroca, Á., Ferrandis, M., Wang, R.: Antiviral properties of alginate-based biomaterials: promising antiviral agents against SARS-CoV-2. ACS Appl. Bio Mater. **4**, 5897–5907 (2021)
32. Cano-Vicent, A., Hashimoto, R., Takayama, K., Serrano-Aroca, Á.: Biocompatible films of calcium alginate inactivate enveloped viruses such as saRS-CoV-2. Polymers (Basel). **14**(7), 1483 (2022)
33. Minsart, M., Van Vlierberghe, S., Dubruel, P., Mignon, A.: Commercial wound dressings for the treatment of exuding wounds: an in-depth physico-chemical comparative study. Burns. Trauma. **10,** tkac024 (2022)
34. Abraham, E., Weber, D.E., Sharon, S., Lapidot, S., Shoseyov, O.: Multifunctional cellulosic scaffolds from modified cellulose nanocrystals. ACS Appl. Mater. Interfaces **9**, 2010–2015 (2017)
35. Danafar, H., Davaran, S., Rostamizadeh, K., Valizadeh, H., Hamidi, M.: Biodegradable m-PEG/PCL core-shell micelles: preparation and characterization as a sustained release formulation for curcumin. Adv. Pharm. Bull. **4**(2), 501–510 (2014)
36. Wang, Q., Zhang, J., Wang, A.: Preparation and characterization of a novel pH-sensitive chitosan-g-poly (acrylic acid)/attapulgite/sodium alginate composite hydrogel bead for controlled release of diclofenac sodium. Carbohyd. Polym. **78**, 731–737 (2009)
37. Sudhakar, Y., Kuotsu, K., Bandyopadhyay, A.K.: Buccal bioadhesive drug delivery—a promising option for orally less efficient drugs. J. Control. Release. **114**(1),15–40 (2006)
38. Lee, K.Y., Mooney, D.J.: Alginate: properties and biomedical applications. Prog. Polym. Sci. **37**(1), 106–126 (2012)
39. Tan, H., Marra, K.G.: Injectable, biodegradable hydrogels for tissue engineering applications. Materials (Basel). **3**(3), 1746–1767 (2010)
40. Malinowska, P.: Algae extracts as active cosmetic ingredients. Zeszyty Naukowe / Uniwersytet Ekonomiczny w Poznaniu. nr **212,** 123–129 (2011)
41. Fabrowska, J., Łęska, B., Schroeder, G., Messyasz, B., Pikosz, M.: Biomass and extracts of algae as material for cosmetics. In: Kim, S.K., and Chojnacka, K. (eds.) Marine Algae Extracts, pp. 681–706, Wiley (2015)
42. Wang, H.M.D., Chen, C.C., Huynh, P., Chang, J.S.: Exploring the potential of using algae in cosmetics. Bioresour. Technol. **84**, 355–362 (2015)
43. Corradini, E., de Moura, M.R., Mattoso, L.H.C.: A preliminary study of the incorparation of NPK fertilizer into chitosan nanoparticles. Express Polym. Lett. **4**, 509–515 (2010)

44. Guilherme, M.R., Aouada, F.A., Fajardo, A.R., Martins, A.F., Paulino, A.T., Davi, M.F.T., Rubira, A.F., Muniz, E.C.: **72**, 385 (2015)
45. Ahmed Khan, T., Zakaria, M.E.T., Kim, H.-J., Ghazali, S., Jamari, S.S.: Carbonaceous microsphere-based superabsorbent polymer as filler for coating of NPK fertilizer: fabrication, properties, swelling, and nitrogen release characteristics. J. Appl. Polym. Sci. **137**, 48396 (2020)
46. Maity, C., Das, N.: Alginate-based smart materials and their application: recent advances and perspectives. Top. Curr. Chem. **380**, 3 (2021)
47. Pawar, S.N., Edgar, K.J.: Alginate derivatization: a review of chemistry, properties and applications. Biomaterials. **33**(11), 3279–3305 (2012)
48. Ma, X., Wen, G.: Development history and synthesis of super-absorbent polymers: a review. J. Polym. Res. **27**, 136 (2020)

Starch-Based Superabsorbent Polymer

Jaylalita Jyotish, Rozalin Nayak, Debajani Tripathy, Srikanta Moharana, and R. N. Mahaling

Abstract Superabsorbent polymers (SAP) were made by cross-linking carboxymethyl cellulose (CMC) and starch, and their efficacy as a water-retaining aid for irrigation was evaluated. For maximum water retention, aluminium sulphate octadecahydrate was used in the cross-linking of the SAP. For the polymer structure, we chose biodegradable and renewable materials like vegetable starch and chemically modified cellulose fibres. Biodegradable polymers can be easily synthesized from carbohydrates since they are both abundant and renewable. For their unique combination of biocompatibility, biodegradability, renewability, and non-toxicity, starch-based SAPs stand out among the others. In addition, starch is cheap and abundant. Alkaline saponification and graft copolymerization with an acrylic monomer and a free radical initiator were used to transform cassava starch into a semi-synthetic superabsorbent polymer. In this chapter, we explored the synthesis and properties of starch-based superabsorbent polymers as well as their applications.

Keywords Starch · Superabsorbent · Polymer

1 Introduction

Superabsorbent polymers (SAPs) are cross-linked hydrophilic 3D macromolecular networks, which have the super capacity of absorbing and retaining significant amounts of liquids [1, 2]. The absorption of water and related liquids by SAPs depends on the hydrophilic functional groups (–OH, $–NH_2$, SO_3H), degree of cross-linking and porosity. Commercialized SAPs can retain about 100–1000 times of liquid per

J. Jyotish · R. Nayak · R. N. Mahaling (✉)
Laboratory of Polymeric and Materials Chemistry, School of Chemistry, Sambalpur University, Jyoti Vihar, Burla 768019, Odisha, India
e-mail: rnmahaling@suniv.ac.in

D. Tripathy · S. Moharana (✉)
School of Applied Sciences, Centurion University of Technology and Management, R.Sitapur, Paralakhemundi, Odisha, India
e-mail: srikantanit@gmail.com

S. Pradhan and S. Mohanty (eds.), *Bio-based Superabsorbents*, Engineering Materials,
https://doi.org/10.1007/978-981-99-3094-4_7

its own weight [3]. A hydrogel known as the superabsorbent polymer (SAP) is a material that is capable of soaking in and storing enormous amounts of water or other aqueous solutions [4]. In general, SAPs may be divided into two categories: synthetic (based on petrochemicals) and naturally occurring (e.g., polypeptide and polysaccharide based). The vast majority of today's superabsorbents are produced by solution or inverse-suspension polymerization techniques using acryl amide (AM), acrylic acid, and its salt as the starting materials. Many people's attention has recently been drawn to bio-based SAPs as a result of the decreased environmental effect that they have. These have found their primary application in the manufacture of sanitary goods, most notably disposable napkins and diapers, in which they serve the purpose of absorbing bodily fluids such as blood and urine. Granular forms of agricultural grades of superabsorbent polymers are utilised in dry regions for the purpose of preventing the loss of soil moisture [1].

Now-a-days, the superabsorbent polymer has received immense attention due to its distinctive 3D network structure, presence of functional groups (e.g., hydroxyl groups, amino groups, sulphonic acid groups and carboxyl groups) [5]. The inherent advantages of SAPs over more traditional water-absorbing materials like cotton, sponge, and pulp make them a promising candidate for use in a wide variety of applications. These include soil conditioners, slow-release fertilisers, drug delivery systems, coal dewatering materials, hygienic products, and self-healing cementitious materials [6]. Despite these capabilities, the SAPs synthesized from primarily synthetic polymers such as petroleum-based materials cause secondary environmental pollution due to their incompatibility, non-biodegradability and toxic nature [7]. Because SAPs can be broken down into their component parts relatively simply, switching to "greener SAPs" might help alleviate the severity of this sort of issue. Therefore, the design of biodegradable SAPs based on biopolymers such as starch has become a hot research area in recent times due to their availability, biodegradability, biocompatibility and renewability. Superabsorbent polymers are capable of swelling and holding possibly huge amounts of water in the swollen state which release the water and associated ingredients slowly in dry circumstances [8]. Superabsorbent polymers (SAPs) are mainly used for agriculture and healthcare applications where water absorbency and retention are essential [9]. Although the petro-based superabsorbents have high liquid-absorbing capacities, those prepared from natural polymers have a greater world-wide demand in industries due to their low cost, abundance, and also because of the presence of large content of natural polymer renders them biodegradable and hence those become environmental-friendly [10]. Superabsorbent polymers (SAP), a special type of hydrogel, are 3D cross-linked or uncross-linked polymers networks. They may absorb many times their own weight in water without significantly dissolving in the liquid because of their adaptability [11].

Since the turn of the last century, variations in the weather caused by natural causes and global warming have contributed to a rise in the frequency of droughts. Therefore, the application of superabsorbent polymers (SAPs) to the soil that has been harmed can be one approach that is practical for the rehabilitation of degraded lands. These are a unique form of polymeric material that have a three-dimensional network that is only weakly cross-linked and has the ability to absorb and store a significant quantity of water and other liquids [12]. The SAP structures are neutral or ionic repeat units that produce an osmotic pressure difference between the water medium and the structure of the SAP to the water molecules to diffuse into the structure [13]. In this chapter we have studied the latest overview of Superabsorbent polymers, production, characteristics and uses of several starch based superabsorbent polymers.

2 Overview of Superabsorbent Polymer

Generally, SAPs based on petroleum polymers such as acrylamide, acrylic acid or methacrylic acids are not biodegradable, and the disposal of their waste products creates severe environmental problems, and hence requires further treatments. Therefore, SAPs based on natural polymers as better substitute for the conventional petro-polymers that are actively receiving attention due to their biodegradability, biocompatibility and non-toxicity [14]. SAPs can absorb and preserve aqueous fluids up to 100 times their weight through H-bonding under pressure. Once absorbed, those release certain amounts of that liquid subsequently depending on their environment. The absorbed fluid is subsequently stored in the multiple tiny holes present in the molecular polymeric network of SAP and then the material creates a gel-like substance which locks the absorbed fluid in. Generally, the cross-linked SAPs of low-density can swell to a greater extent and have a higher absorbency. The water absorption capacity of SAPs also varies with the ionic concentration of the aqueous solution [9]. Superabsorbents or hydrogels can absorb excess amount of water in which the absorbed water is hardly releasable even under some pressure. Superabsorbents have raised considerable researches and have been used in agriculture, health and horticulture since the first SAP was detailed by the US Department of Agriculture in 1969 due to their excellent characteristics [15]. However, petroleum based SAPs are gradually replacing with the bio-based SAPs as the petro-based SAPs lead to serious environmental concerns [16]. Currently, these are the most desirable products for several potential applications in various diverse fields such as agriculture, drug delivery, chemical industry, biosensors, thermal energy storage, hygienic products and tissue engineering [17, 18]. SAP has an ability to hold a large amount of water to form hydrogel [19, 20] and has been widely used in agriculture [21], medical materials [22] waste-water treatment [23] and food science [24]. SAP absorbs aqueous solutions through H-bonding with the water molecules. The presence of sufficient hydrophilic functional groups on the polymer backbone of SAP makes the liquid adsorption ability excellent [25–27]. In recent years, many new techniques and processes have been developed for the production of high-performing superabsorbent polymers using natural polymers as raw materials with the rapid progress in material science [28, 29]. Now-a-days, it has become a great challenge to synthesize bio-based superabsorbent polymers to address the environmental concerns [8]. These cross-linked hydrophilic polymers have the ability to absorb and retain a significant amount of water or organic solvents into their 3D networks. Recent studies discovered that the natural polymers such as starch could be used for the synthesis of superabsorbent polymers. Starch-based superabsorbent polymer (SBSAP) has been receiving increasing attention due to its relative low production cost, high water or aqueous solution absorbing and capturing capacity and considerable applications in diverse fields. In addition, these have attracted wide scientific and industrial interest for use as better substitutes to the synthetic superabsorbent polymers [10]. Figure 1 shows the chemical structure of the starch-based superabsorbent polymer [20]. SAPs have attracted great attention due to the wide range of potential application provided by them. Now-a-days, SAP has received significant research attention because of the superior properties offered by them. To control the environmental issues, growing efforts have been given for the development of "greener" or bio-based SAPs from natural polymers like starch which are not only sustainable and widely available, but also can be biocompatible and biodegradable [30]. Bio-based superabsorbent is an interesting polymeric material having excellent water absorption capacity which can act as a soil conditioner and an active agent carrier along with others. However, extensive usage of SAP derived from conventional petro-based polymers is creating serious threats to the sustainability of the ecosystem [31].

Since starch is one of the best natural polymers with multiple excellent features, it has been widely used to replace the more traditional petro-based or synthetic polymer. This is one of the main factors influencing the scientific and commercial communities' interest in starch-based superabsorbent polymers (SBSAPs). Recent years have seen a rise in the popularity of bio-based SAPs due to the novel and appealing combination of their properties. These properties include an easy availability, excellent mechanical properties, surface chemistry, biodegradability, biocompatibility, and a wide range of applications in a variety of fields. In contrast to the typical SAPs that are derived from petroleum, the starch-based SAP, also known as SBSAP, has

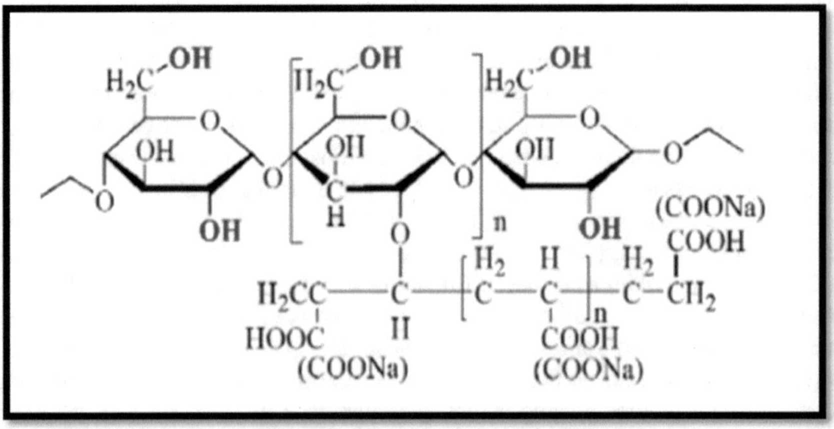

Fig. 1 Starch based superabsorbent polymer [20]

been proven to be better due to its increased biodegradability, biocompatibility, abundance, and low cost [32, 33]. Because of the significant environmental and allergy issues caused by the previously employed non-biodegradable materials [34, 35], the use of bio-based SAPs for the purpose of maintaining hygiene has garnered a great deal of interest in recent times. Due to the fact that it has the potential to be used in a variety of diverse disciplines, the superabsorbent polymer, often known as SAP, has earned a respectable standing in the scientific and industrial communities during the past two to three decades [31]. Because of their high capacity for water absorption and exceptional network structure, they have applications in a wide variety of industries, including agriculture, wound dressing, medication delivery, tissue engineering, waste-water treatment, personal hygiene, and many others [36]. Both academic institutions and commercial businesses are beginning to take an interest in biodegradable polymers [1]. In addition, natural polymers have garnered a lot of attention due to the fact that they are biodegradable, biocompatible, inexpensive, and abundant. Amylose, which is a linear structure, and amylopectin, which is a branching structure, are two examples of the types of microstructure that may be found in starch [37]. Because of how easily it can be produced in abundance and how easily it can be broken down, it is thought to be one of the finest bio-polymers for the production of encapsulating materials. It is suitable for the production of cross linking polymers due to the high amount of hydroxyl groups (–OH) that it contains. After polymerization, the film-forming characteristics of starch can be improved with the addition of a biodegradable and biocompatible polymer, such as poly (vinyl alcohol) (PVA). [38] Fillers are necessary to increase the mechanical strength of starch/PVA films since these films have many surface-hydroxyl groups (–OH), high surface activity, inadequate mechanical characteristics, and an excessive amount of hydrophilicity [39].

Starch is a naturally occurring biopolymer that possesses high hydroxyl group content [40]. Because of its biodegradability, hydration-swelling feature, low cost, and abundant availability [41], it has seen widespread application in the synthesis of green SAPs. Unfortunately, the shorter service life that results from the increased biodegradability of starch-based SAPs (SBSAPs) makes them less desirable for use in agricultural applications. It is possible to modify starch with certain antibacterial nanoparticles, such as zinc oxide (ZnO) and titanium dioxide (TiO_2) [12], which would increase the starch-resistance in an environment including dirt. The most common foods that contain starch are maize, potatoes, and wheat. Starch is made up of glucose units that are joined to one another by glycosidic bonds and is made up of amylose and amylopectin. Applications can be found for it in the production of biofuels as well as alcohols. Additionally, it can be utilised as a thickening or bonding agent. The ratio of amylose to amylopectin has a significant impact on the properties that are shown by the starch. When there is a bigger quantity of amylopectin, the viscosity increases, which in turn decreases the mobility of the chains. If the amylose concentration is larger, the swelling capacity will be greater, and the grafting efficiency will be improved [1]. In order to improve the efficiency of the fertiliser, double-coated polymers consisting of a starch-based SAP as the outer coating and ethyl cellulose as the inner coating have already been created as slow-release fertilizers. These polymers are intended to be used as slow-release fertilizers. Metals can also be absorbed via starch-g-poly(acrylic acid)-sodium humate or starch-graft-acrylic acid montmorillonite SAPs [42]. Both of these SAPs can be found in starch. It is possible to use starch-graft-poly(acrylamide) SAPs for soil conditioning treatments in order to successfully absorb soil moisture [43, 44]. Carboxymethyl starch-g-polyacrylamide is utilised in the production of slow-release fertilizers. Starch is an abundant renewable polysaccharide which may be manufactured at a lowest price and at vast sizes. According to the weight, this is composed of 75–80% amylopectin and 20–25% amylose [45].

The absorptive properties of starch-based superabsorbent polymers allow them to function as reservoirs in agricultural fields when they are placed there. These reservoirs can hold an excessive quantity of water and nutrients. And when the periods of deficit arrive, those SAPs are able to release the water and nutrients that they had previously saved in order to promote the growth of the plant [29]. Due to the water-holding ability of SAP, its application in agricultural areas results to a rise in the percentage of crops that survive as well as a reduction in the frequency of irrigation [20].

3 Preparation of Starch Based Superabsorbent Polymer

At high starch concentrations, a lot of work has been put into developing new methods that can prepare starch-based superabsorbent polymers in a way that is both kind to the environment and cost-effective. By measuring the torque and temperature, a newly designed twin-rotor mixer that has better sealing to produce an oxygen-free environment may be utilised to examine the physical and chemical changes that occur throughout the melting process. A HAAKE rheometer that incorporates a twin-rotor mixer is employed, and this rheometer was changed to increase its feeding and sealing, as well as to produce an atmosphere that is oxygen-proof [46]. This method of mixing in one step may successfully handle very viscous materials, and it also enables the synthesis process to be completed in one sequential step with a significant reduction in the amount of time required. Prior to it, a process known as reactive extrusion (REX) had been developed for the modification of starch beginning in the early 1990s [47–49]. Extrusion, on the other hand, is a complicated process that makes it difficult to manipulate the qualities of the end products and to predict correctly how the chemical reaction will play out. In addition to this, the financial investment required for the necessary equipment is substantial. A mixer, on the other hand, is significantly more affordable and has less facility, but it is still very important in the field of polymer and plastic engineering. This includes the blending and processing of thermoplastics, as well as the vulcanization and mastication of thermosets. As a result, this mixer/blender type reactor has garnered a significant amount of interest in recent times as a result of the efficacy with which it can manufacture materials with acceptable performance characteristics.

However, there has not been a comprehensive investigation of the connection between the material structure, reaction circumstances, and the long-term release behavior of fertilizer-filled SBSAP that was produced using a reactive mixing technique. On the other hand, one can observe the effects of the cross-linker (N,N′-MBA, also known as N,N′-methylene-bisacrylamide), the saponification agent (NaOH), and the initiator (CAN, also known as ceric ammonium nitrate) on the characteristics of SBSAP under various reactive mixing conditions (temperature, time, and shear intensity). There is a great deal of published material accessible on the subject of the production of bio-based SAPs. These have undergone a chemical transformation in order to lessen their toxicity and enhance their biocompatibility. In order to understand the effect of the original molecular structure on the release performance and characteristics of SBSAP, three distinct numbers of the same starch were collected, each of which had a different ratio of amylose to amylopectin. Therefore, the reliance of slow-release performance of SBSAP (up to 45 days) on the structure, swelling degree, and gel strength of SBSAP was demonstrated by producing with the use of a reactive mixing procedure [8]. This was done in order to establish the relationship between the three variables.

A reactor system that is capable of giving high torque for the processing of very viscous materials was developed by combining a HAAKE Rhemix 600p twin-rotor mixer (as shown in Fig. 2) with a HAAKE Rheocord Polylab RC 500p system. This combination resulted in the creation of the system. Because the starch modification is required, beginning in an oxygen-free environment and be carried out with water, the mixer had to be maintained shut at all times so that it could be completed. Teflon film was employed to seal the spaces between the three barrels of the mixer, and a silicon-rubber cover was utilised to seal the feeder on top. Both of these measures were taken to avoid the loss of moisture while the reaction was taking place. On the silicon rubber cover [8] there is a connecting tube for emptying the reaction chamber with nitrogen and a funnel for liquid chemicals. Both of these features are located on the cover.

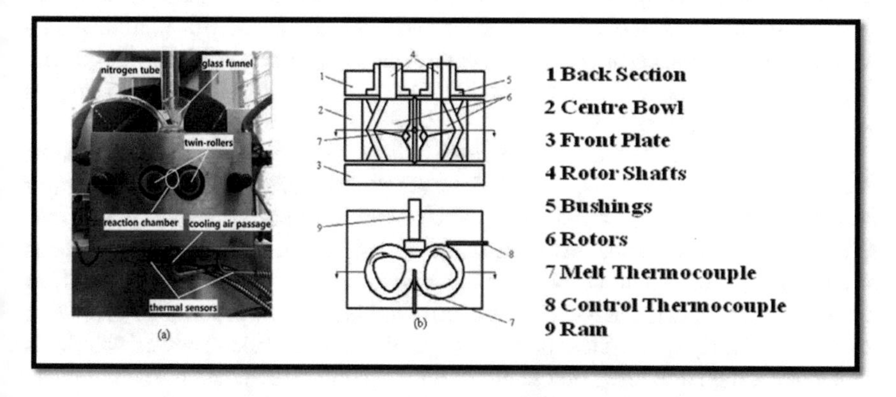

Fig. 2 **a** Photo and **b** schematic of the modified HAAKE Rheomix mixer [8]

3.1 Preparation of Urea-Embedded SBSAPs

There have been two approaches taken in order to manufacture urea-embedded SBSAPs. In Method 1, the modification of the starch is completed first in order to produce the SBSAP. However, urea is added to the mixer after the mixer has been used to incorporate the urea. In order to obtain a starch-based hydrogel, the starch is grafted with acrylamide at the outset, and then it is cross linked using the cross linker N,N′-MBA. At a temperature of 800 °C, suitable quantities of starch, AM, and distilled water are mixed with N, N′-MBA in a HAAKE twin-roller rotor mixer. After that, the mixing process is started at a speed of 80 °C revolutions per minute for approximately 10 min in order to gelatinize the starch with strong shear stress. After that, the temperature of the mixer is brought down to 650 °C by using compressed air, and then the HAAKE mixer begins to receive nitrogen. After the reaction chamber of the instrument has been purged with nitrogen for ten minutes, a predetermined quantity of CAN is then injected to the mixer while it is operating at 650 °F with 80 revolutions per minute. This operation begins the grafting process. After the initial 10 min of reaction, a solution of NaOH is added to the mixer, and the process of saponification is then continued for an additional 10 min at 80 revolutions per minute at 650 °C. Following the completion of the saponification process, urea is added to the mixer, and the mixing process is continued for an additional 40 min at 80 revolutions per minute and 800 °C. "SBSAP-urea-M1" is the name given to the SBSAP that has been implanted with urea through the use of Method 1.

Method 2: This method entails only one step, and it allows for the manufacture of urea-embedded SBSAPs by the reaction of starch and its subsequent blending with urea. Previously, the granule urea is ground into a powder and then combined with starch. The same procedure as the initial method is executed in its entirety. The sample resulting from this technique is designated as "SBSAP-urea-M2."

The preparation of the neat SBSAP (without urea) also involves using a technique that is analogous to Method 2 for the chemical modification of starch; with the exception that urea is not added into the process. After the reactive mixing is complete, the components soak in distilled water for two days to expand. This is followed by washing the components in distilled water until the samples become fully transparent. After drying the samples in an oven at 500 °C to a consistent weight for one day, the samples are then pulverised.

For the most part, starch is incorporated into synthetic polymers as a copolymer. Before the copolymerization, the solutions are pregelatinized using microwaves or heat to create a more uniform network. The starch-SAPs used for slow-release urea fertilizer in agriculture undergo a single melting stage (as shown in Fig. 3). The effects of amylase concentration, CAN (ceric ammonium nitrate) as an initiator, NaOH as a saponification agent, and N-MBA as a cross-linking agent were studied in relation to temperature, time, and shear rate. Methods of characterization showed that SAPs with either a high amylose (Gelose 50) or amylopectin content had been effectively produced. Porous crosslinked SAPs gels with a high amylase concentration were shown to have a much higher storage modulus, as measured by rheometry. SAPs

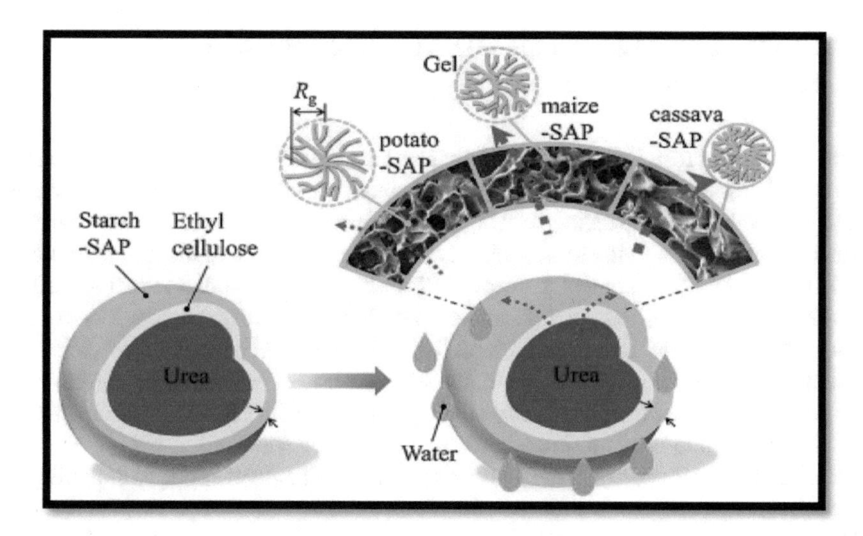

Fig. 3 Schematic representation for the relationship between starch-SAPs and the slow release behavior of double coated fertilizer [50]

microstructure, gel strength, and WAC all affect how quickly they release urea. Traditional fertilizer phosphate rock (PHR) is poorly absorbed by plants. Therefore, SAPs based on the sulphonated corn starch (SCS)-pAA embedded phosphate rock (SCS-pAA-PHR) was produced to enhance the usage of the PHR, to regulate the release of fertilisers, and to include the water catching capacity. The SCS-pAA-PHR WAC was proportional to the % of AA that was neutralized in the polymeric network. As the synthetic SAHs' PHR content rose, their WAC fell. And the SAHs have great nutrient-releasing properties for plants. Soil amendment humus (SAH) made from cassava starch may be used to store water for plant roots and improve the soil quality. After SAP was applied, the soil's density fell but the porosity and WAC both rose. To increase the fertilizer's availability, scientists created a double-coated sustained release fertiliser using starch-SAPs on the outside and ethyl cellulose on the inside. The relationship between the properties of Starch-based SAP and the slow release feature of double-coated fertiliser is depicted schematically in Fig. 2. The WAC was helped along by the nano-sized gel's enhanced sustained releasing-property and decreased absorption rate. Based on an analysis of SEM pictures, we may deduce that potatoes, followed by maize and then cassava, have the highest WAC. The SAP made from potato starch exhibited the smallest grid size, relatively loose fractal gels, the lowest water diffusion rate, and the biggest WAC, resulting in a fertilizer with the best delayed releasing feature [50].

4 Cross Linker Effects on SBSAP Characteristics and Microstructure

The amount of cross linker (N,N′-methylene-bisacrylamide) affects the absorbent, micro-structural and rheological features of starch-based superabsorbent polymers prepared by the one-step method at a high concentration of starch (0.27:1 w/w starch–water). The increase in the cross linker amount seemingly alters the microstructure and absorbent and rheological features of SBSAPs. Previous studies reveal the influence of reaction conditions such as amylase-amylopectin ratio on the properties (e.g., water absorption rate) and the structural features for SBSAPs prepared by the reactive mixing method and solution methods. Also, the cross linking density can change the porosity and the network structure of SAPs (as shown in Fig. 4a) prepared at low starch concentrations and ultimately the absorption behaviors [51].

In fact, the chains between the junctions points dominantly respond to the external stress by chain compression for the cross-linked polymer [52]. The swollen starch-SAPs contain mainly two kinds of "chains in-between", counting the PAM chains with one end linked to the MBA cross linker and the other end to –OH on anhydroglucose units and the starch chain segments with the two-end graft by PAM chains. The higher amount of MBA topologically divides the large starch chains into smaller fragments and increases the ratio grafted starch carbons with PAM (known by the larger GR). These shortened chain segments and the retrenched PAM chain length (i.e., smaller L_{PAM}) display more rigidity than the longer counterparts and ultimately display the storage modulus and robustness for the SBSAP gel matrixes. Moreover, the less stretched SBSAP chains and the increased cross-linking density also contribute to enhance the resistance of gel matrixes to external stress (the higher

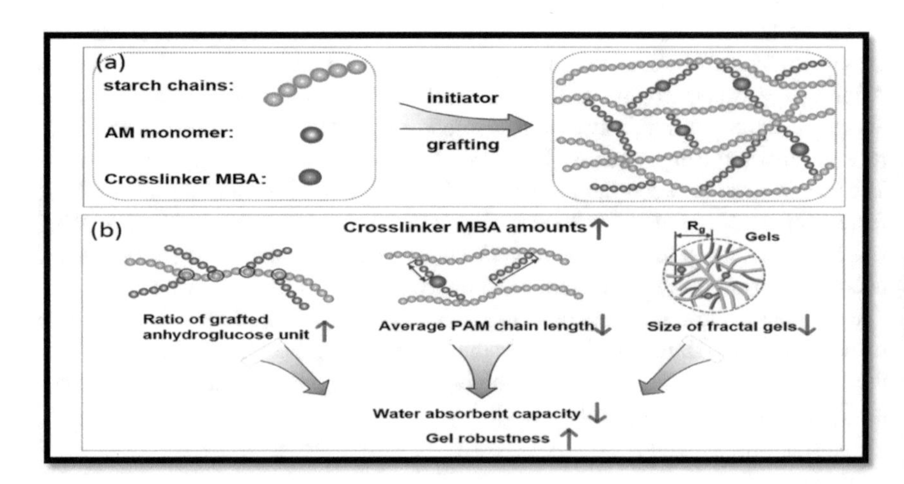

Fig. 4 Schematic representation for **a** formation of 3D network of starch-based SAP and **b** cross linker amount affects the rheological features and water absorbance property of starch-based SAP [6]

storage modulus and robustness). The use of more MBA cross-linker leads to shortened PAM chains, and increased ratio of starch carbon grafted the weakened stretch of molecule chains and the risen cross-linking density. These tends to reduce the flexibility of the molecule chains within the SBSAP gel, thus resulting in lessening the mobility of the chains (a lower strain under equal external stress). As a result, the mechanical energy dissipation of SBSAP chains at the time of testing displays a downtrend with the increase of MBA amount. Note that the micro-structural features can be altered with the increase in MBA amount and thus the absorbent and rheological features of SBSAPs. Then, the evolutions in the SBSAP properties have been discussed from a microstructure viewpoint. Other than the increment of cross-linking density, the greater quantity of MBA allows the shortened average length of PAM chains, and the raised ratio (GR) of starch carbons grafted by PAM. Such structural properties certainly stifle the chain stretch within the SBSAP mass fractal gels, and contribute to the formation of SBSAP chain networks of reduced size; which disables the holding and absorbing events of water molecules into the molecular chain networks of SBSAP (as shown in Fig. 4b). Then, a reduction in the water absorbent capacity occurs with the increase in the MBA amount. On the other hand, the mobility and flexibility of the chain segments within the SBSAP gel goes on decreasing due to the reduced L_{PAM} and GR, with elevated cross linking density and lowered chain extension. Moreover, the SBSAP gels show enhanced storage modulus and robustness as well as the lowered energy dissipation (reflected by the loss factor) of the chains. So, the SBSAPs having higher water absorption capacity display lower robustness and storage modulus, specifying a lesser ability to retain water for those SAPs related to their comparably low cross linking density [6].

5 Graft Copolymerization

Lanthong and his co-workers have synthesized biodegradable superabsorbent polymers by the graft copolymerization of acrylamide (AM)-itaconic acid (IA) onto starch by a redox initiator system of N,N,N′,N′-tetramethylethylenediamine (TEMED) and ammonium persulphate (APS), in the presence of the cross linking agent, N,N′-methylene bis-acrylamide (N-MBA), sodium bicarbonate foaming agent, a foam stabilizer. Hydrogel or SAP is a principal type of partially cross linked polymer materials having the capacity of absorbing large amount of fluids, especially water. Natural polymer like starch, have been used in the preparation of hydrogels or SAPs. Starch-based SAPs are developed by grafting the starch with acrylonitrile, acrylamide, acrylic monomer and ε-caprolactone [53]. Those can be prepared by the chemical copolymerization [54] and with the use of γ-ray irradiation [55, 56]. SAP or hydrogel transition (changes in volume) occurs due to the changes in the environmental conditions. Therefore, the behavior of a highly swollen SAP is a function of molecular network characteristics, such as diffusion parameters, degree of swelling, pH, cross linking density and so on [57]. Itaconic acid and acrylamide are water-soluble and can allow polymer chains with the carboxylic groups (diprotic acid) and amide, respectively. The side chains may raise the hydrophilic property of the copolymer. They prepared starch-based SAP by grafting poly [acrylamide-co-(itaconic acid)] on starch, where the ratio of starch and poly [acrylamide-co-(itaconic acid)] is in the range of 1:4–2:1. Their absorption capacity varies with the amount of starch and itaconic acid content.

In the starch graft copolymerization, the starch was gelatinized at first. Then, the gelatinized starch was cooled to 45 °C. After that, measured amount of acrylamide (AM), itaconic acid (IA), N-MBA, ammonium persulphate (APS), $NaHCO_3$ and TEMED were added into the reaction under stirring to get a homogeneous solution. Then, a semi-viscous solution appeared after stirring the reaction mixture for 30 min under the nitrogen atmosphere. The founded product was precipitated by methanol and then dried at 65 °C. So, the bio-based SAPs were formed by the graft copolymerization of the gelatinised starch and the acrylamide-itaconic acid by foamed solution polymerisation utilising TEMED and APS as a co-initiator and oxidation–reduction initiator, respectively, while N-MBA acts as the cross linker. The presence of both the itaconic acid and acrylamide is important for the grafting reaction on starch to get high absorbency. The water absorbency can be calculated as,

$$\text{Water absorbency } (Q) = (W_1 - W_0) / W_0$$

where W_1 is the weight of water swollen gel (g) and W_0 is the weight of the dry polymer (g).

The weight of polymer in grafts and starch (substrate) is regarded as the percentage (%) of grafting ratio which is calculated as,

$$\% \text{ grafting ratio} = \text{Weight of the starch grafted polymer } \times 100/\text{Weight of starch}$$

The % of total synthetic polymer formed which had been grafted to the starch is known as the "percentage grafting efficiency" which is calculated as,

$$\begin{aligned} \% \text{ Grafting efficiency} &= \text{Weight of starch grafted polymer} \\ &\times 100/\text{Weight of free polymers} \\ &+ \text{Weight of polymer grafted} \end{aligned}$$

The percentages grafting ratio and grafting efficiency are enhanced by increasing the cross linker concentration in graft copolymerization. Moreover, the water absorption of the graft copolymer decreased significantly by increasing the ionic strength of salt solutions. In addition, constant water absorption is observed while immerging the anionic SAP in a high pH buffered solution [53].

6 Effects of Amylase-Amylopectin Ratio on Starch Based Superabsorbent Polymers

Bio-based superabsorbent polymers (SAPs) are synthesized by grafting the acrylamide into starches and then cross linking with the cross linker, N,N-methylene-bisacrylamide. This work especially focused on the effects of the amylase-amylopectin ratio of starches from a similar source on the grafting reactions and performance of the resultant SBSAPs. The acrylamide groups grafted on to starch are analyzed by FTIR to characterize each superabsorbent polymer. Grafting efficiency and grafting ratio are calculated by using a gravimetric method and the length of the grafted segment and the graft position are detected by NMR. The graft reactions, relationships between the microstructures of starches and the performance of the SAPs are investigated based on the amylose content in starches. Under the same reaction conditions, the grafting efficiency and the grafting ratio increased with the increase in the amylose content, which coordinates with the water absorption ratio. It is known from the NMR results that the acrylamide group is generally grafted on to C6, and the length of the grafted segment goes on decreasing with the increase of amylopectin content normally, and particularly for waxy starch. Branched structure and the high molecular weight of amylopectin decrease the mobility of polymer chains and rise viscosity, which can explain the grafting reactions and the performance of the SBSAPs [1].

SBSAPs are also formed by grafting the starch with the acrylic monomer, e-caprolactone and acrylamide [58–60]. These can be developed by the chemical copolymerization [61]. The first SBSAP was formed through the base hydrolysis of starch-graft-polyacrylonitrile [62] and the subsequent SBSAPs have been prepared by grafting the starch with the acrylic monomer, acrylamide (AM) and caprolactone. These can also be prepared by chemical co-polymerization [58–60] and the utilization of gamma-ray [63]. Normally, a polysaccharide reacts with an initiator in either of the two processes in graft copolymerization. Firstly, the initiator (generally Ce^{4+}) and the neighbouring –OH groups of the saccharide units will interact to form the redox pair-based complexes. And, these complexes then dissociate to form carbon radicals on the polysaccharide substrate through homogeneous cleavage of the saccharide C–C bonds. The graft copolymerisation of vinyl monomers and crosslinker is initiated by those free radicals on the substrate. On the other hand, an initiator like persulphate may extract hydrogen radicals from –OH groups of the polysaccharide to generate the initiating radicals on the polysaccharide backbone. Several starches have been employed to prepare SBSAPs, such as corn [64], potato [65] and cassava [66]. From these, corn starch and its derivatives are much popular base materials, and are made use to prepare SBSAPs [67–70]. Comer and Jessop [71] investigated the grafting efficiency of starch-g-poly (methyl methacrylate) synthesized by emulsion photo-polymerization without a photo-initiator utilizing two types of starch (high amylose maize starch and high amylopectin waxy corn starch), and found to know that there were no such major differences. Rath and Singh [72] have reported that the amylopectin-g-PAM (amylopectin-g-polyacrylamide) showed better performance in flocculation than starch-g-PAM and amylose-g-PAM, and that the longer and fewer PAM chains were most effective. But there are no such systematic analyses of the effects of the amylose-amylopectin ratio on the starches from equal source on graft reactions and the performance of SBSAPs. The microstructures of the starch are still being elucidated, yet it is generally known that starch is a heterogeneous material containing two kinds of microstructure: one is linear (amylose) and another one is branched (amylopectin). The linear structure of amylose makes its characteristics more closely resemble to that of the conventional synthetic polymers and the molecular weight of amylose is about 10 times higher than that of the conventional synthetic polymers. And amylopectin, on the other hand, is a branched polymer, molecular weight is much higher than the amylose. Branched structure and the high molecular weight of amylopectin decrease the mobility of the polymer chains. Moreover, most native starches are having a crystallinity of about 15–45% i.e., semi-crystalline [71]. The branching points of amylopectin and amylose form the amorphous regions. The less branching chains in amylopectin are the chief crystalline components in the granular starch [72]. Starches having different amylose/amylopectin ratios have the different rheological properties [73] and phase transition characteristics [74]. Corn starches of different amylose contents (maize, 29.0%; waxy, 4.3%; Gelose 50, 50% and Gelose 80, 80%) have been used as model materials. The SBSAPs are synthesized by grafting the AM on to corn starch employing ceric ammonium nitrate (CAN) as an initiator and N,N-methylenebisacrylamide (N-MBA) as the cross linker. The relationships between grafting reactions and the microstructures of the starches, and performance of SAPs are analyzed based on the amylose content, grafting efficiency, grafting ratio and the water absorption.

Figure 5 represents the LOS plots, water absorption curves and their fit curves for the starch-based SAPs. The fitted amounts of water absorbed at the swelling equilibrium state were in agreement with the WAC values. This revealed that the water absorption of starch-based SAPs went through two consecutive phases, i.e., a typical dual-phase process, with the first phase having a higher rate than the second. In fact, starch-based SAPs had polymer networks constituted by PAM/anhydroglucose chains. The highly hydrophilic groups especially on the PAM chains could strongly interact with the water molecules while being immersed in water. This not only induces a swelling of the polymer networks with chain relaxation but also promotes the water diffusion towards the molecular network structures within the starch-based SAPs. As the water absorption proceeded, a second absorption phase emerged, where the chain relaxation became an important event, but water diffusion was weakened. In the earlier phase, the order of absorption rate was as, G80-SAP < G50-SAP < maize-SAP < waxy-SAP, which was equal to the decreasing trend of LPAM. The absorption process of starch-based SAPs consists of relaxation of molecular chains in the network and diffusion of water into the polymer network. The lower LPAM (along with a larger GR) led to smaller starch-based SAP chain networks with a higher specific surface area, and thus could enhance the diffusion of water towards to the networks. Also, the relaxation of smaller PAM chains tended to fast reach of an equilibrium swelling state.

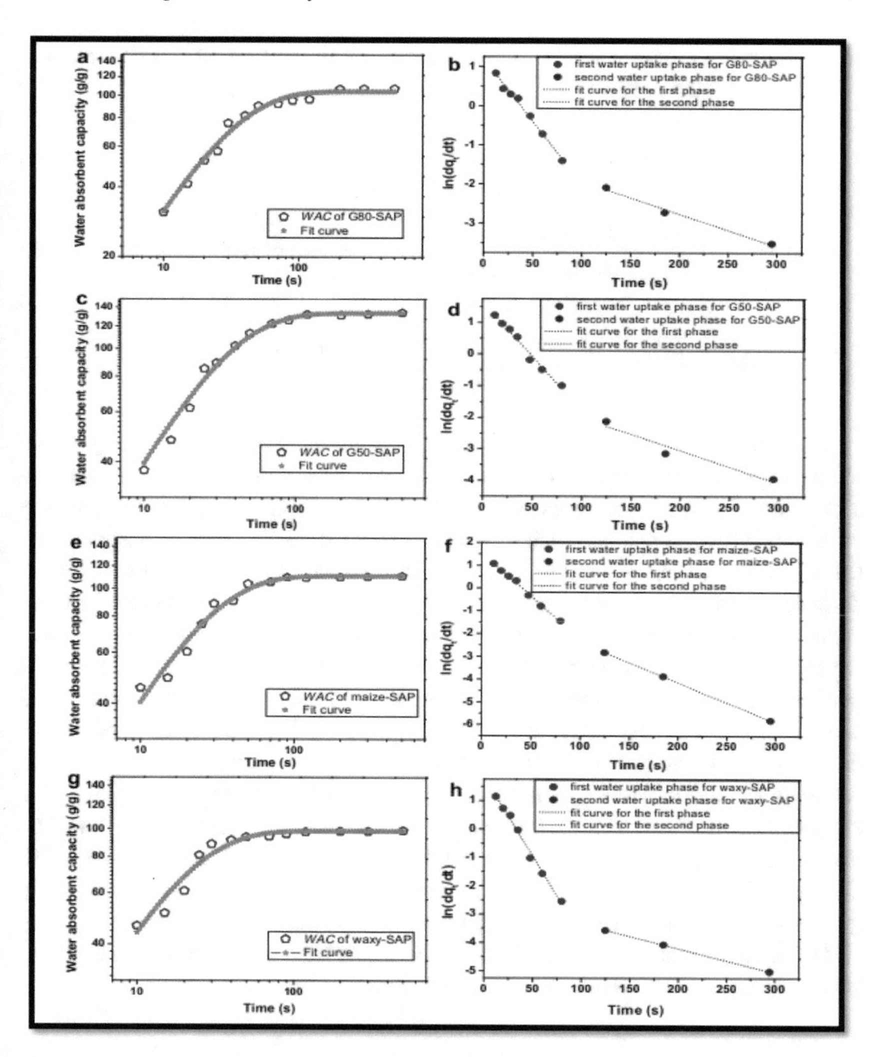

Fig. 5 Typical water absorption curves (**a**, **c**, **e**, and **g**) and LOS plots (**b**, **d**, **f**, and **h**) and their fit curves for G80-SAP (**a** and **b**), G50-SAP (**c** and **d**), maize starch-SAP (**e** and **f**) and waxy starch-SAP (**g** and **h**) [30]

7 Determination of Add-on, Grafting Ratio and Grafting Efficiency

The percentages of free polymer, graft polymer and the soluble starch are determined via separation by washing with distilled water, and acid hydrolysis by using HCl [53]. The % weight of PAM in the graft polymer, or the so-called "% add-on" (AO), can be calculated from the weight difference between the graft polymer and the soluble starch formerly removed by the acid hydrolysis. The grafting ratio (GR), grafting efficiency (GE) and add-on (AO) are calculated as,

$$\mathrm{GR}\,(\%) = \frac{\text{Weight of starch grafted polymer}}{\text{Weight of starch}} \times 100 \tag{1}$$

$$\mathrm{GE}\,(\%) = \frac{\text{Weight of starch grafted polymer}}{\text{Weight of free polymer}} + \text{Weight of polymer grafted} \times 100 \tag{2}$$

$$\mathrm{AO}\,(\%) = \frac{\text{Weight of graft polymer} - \text{Weight of soluble starch}}{\text{Weight of graft polymer}} \times 100 \tag{3}$$

The effects of the amylose/amylopectin ratio on the grafting ratio and grafting efficiency are determined by using a gravimetric method based on the extraction and separation of several substances. The add-on (AO), grafting ratio (GR), and grafting efficiency (GE) are calculated using Eqs. (1)–(3), respectively. It can be observed that the AO, GE, and GR generally increase with the increase in the amylose content. Amylopectin is a branched monomer, and it would be seen that the double helix crystalline structure created by its small branched chains (i.e., only 4–6 glucose), which are not as flexible as the long chains of amylose, is turned apart at the time of gelatinization. Moreover, those short, branched chains stay in a regular manner and keep a certain "memory" after the gelatinization [75]. The small, branched chains create gel-balls, each of which carries the chains from the same sub-main chain. It is known that one amylopectin molecule can form a relatively large superglobe. The slightly tight structure of a gel ball withstands the grafting chemical reaction. The SAP-G50 samples achieve greater values for AO, GE, and GR than the SAP-G80. One feasible explanation for that will be that G80 may not have been completely gelatinized at the first stage of grafting since it has the highest gelatinization temperature. The multi-phase transitions of the G80 at the time of heating, particularly at high temperatures, can hinder the grafting reaction.

8 Effects of the Amylose-Amylopectin Ratio on Water Absorption

The SAPs based on the starches with more amylose content record higher water absorption, which corresponds with their higher grafting efficiency (GE) and higher grafting ratio (GR). The highly branched structure and high molecular weight of amylopectin decrease the mobility of the polymer chains and interfere with any tendency to become closely oriented enough to permit major levels of H-bonding. Moreover, based on the NMR results, the greater grafted segments in high amylose content starch may have also contributed to higher water absorption. SAPs are synthesized by grafting the acrylamide onto starches, and then cross linking with the cross linker, N,N-methylene-bisacrylamide (N-MBA). Polyacrylamide (PAA) is grafted onto all the starches. The results of tests on the effects of the amylose/amylopectin ratios show a parallel pattern for AO, GE, and GR, basically increasing with the increase in the amylose content. From the NMR study, it is known that the acrylamide group is especially grafted onto C6, and the length of the grafted chains decreases with an increase in amylopectin content, specifically for waxy starch. High molecular weight and the branched structure of amylopectin lower the mobility of the polymer chains, resulting in resistant chain growth and higher viscosity. The superabsorbent polymers based on higher amylose content show higher water absorption, which corresponds to their higher grafting efficiency (GE) and grafting ratio (GR). The larger graft segments and the flexible linear chain structure of starches with high amylose content also contribute to this result.

9 Applications

9.1 Use of SBSAP in Agriculture for Slow Release of Fertilizer

A key condition for plant growth is a sufficient water supply. But it is not enough to support the growth of plants only by giving them water because the fertiliser can enhance plant growth. Hence, both the fertiliser and water are two essential things for the plant growth, and it is essential to unite the sufficient supply of water and fertiliser in agriculture, mainly in the areas which are suffering from drought [76, 77]. SAP, which can absorb, hold large amount of water and release them slowly in a dry circumstance, has been used extensively in many fields [78–80]. In agriculture, these have been successfully utilised to decrease the death rate of plants, reduce irrigation frequency, and enhance the output of crops in arid and semi-arid lands [81]. The application of fertilisers, especially nitrogen fertilisers, is the most common and effectively used method to boost crop yield. Nitrogen is the vital nutrition for the plant growth, and the lack of nitrogen has been considered the dominant limiting factor for the yield [82]. It has been reported previously that 40–70% of nitrogen (N) and about 80–90% of phosphorus (P) in fertilisers cannot be utilised by crops due to their high diffusivity to the surrounding environment and high solubility in water [83]. Urea is examined because it contains a high content of nitrogen and is the most commonly used nitrogen fertilizer. However, urea is easy to dissolve and wash out with the flowing water, while it is hard to be fixed by the soil particles because it is a neutral organic molecule. The leaching fertiliser pollutes the surface and ground water, causing eutrophication of lakes and reservoirs and ultimately imposing risks on the ecosystem [84, 85]. However, newly created starch-based superabsorbent polymers for the slow release of the fertiliser can solve these types of problems by providing sufficient nutrients for plants, reducing nutrient loss to the surroundings, and enhancing crop yields [86]. A typical method for the slow release of the fertiliser is to apply a coating. Materials for such coatings include the oil-based polymers (e.g., polyvinyl chloride, polystyrene, polyethylene, polypropylene, and the acrylonitrile-butadiene styrene polymer) [87]. Superabsorbent polymers (SAPs) are suitable for agricultural application as a nutrient carrier and soil conditioner [88]. SBSAPs are comparably superior to traditional SAPs in regards to their abundance, biodegradability, renewability, and low cost [33]. In a previous study, it was observed that when SBSAP was prepared by using the one-step method using reactive mixing for the slow release of the urea fertiliser, the urea content was up to 400% and the release properties mainly depended on the water absorption capacity (WAC) and the gel strength. Using the SBSAP, the slow release of urea into water was executed. From that literature, it was also known that, within 1 day, less than 15% urea was released, the release rate exceeded 80% after 30 days, and the release held out for more than 45 days.

Use of water-soluble fertilisers such as urea, mono-ammonium phosphate (MAP), diammonium phosphate (DAP), etc. causes a flash release of nutrients that cannot fulfil the plant's nutrient demand. Achievable recovery efficiency is only 15–20% for phosphatic (P) and 35–40% for applied nitrogen (N) fertilisers. Additionally, the flash release of nutrients results in nutrient loss, economic losses, and subsequent environmental pollution [89]. The best possible solution to these problems is the employment of controlled-release fertilisers (CRF). The fertiliser that releases the nutrients at a moderate rate for a longer period of time as per the plant's demand is known as CRF. Fertilisers can be regarded as slow release fertilisers (SRF) if their amount of nutrient release is (i) less than 15% in 1 day, (ii) less than 80% in 30 days, and (iii) must release 80% during 30–365 days at 25 °C [39]. However, the term controlled release fertiliser (CRF) and slow release fertiliser (SRF) are often named analogous to each other. The encapsulation of granular fertilisers is the method that is both the most effective and the most prevalent one used in the production of CRF [90]. As the degradability of the encapsulating material is a great cause of concern, more attention has been paid in recent years to biodegradable polymer encapsulations that are compatible with the environment. Also, economically feasible methods are being employed for the production of CRFs [91].

There were three methods for combining the SAP with fertilizers. The first method was a normal co-application of SAP with the fertilizers. However, in this case, after fertilization, the electrolyte concentration rapidly increased in the soil solution, and the hydrophilic groups were easily complexed inside the SAP by the excess salt ions, which resulted in a maximum reduction of water absorption. In second method, SAP was used as a carrier where the core fertilizer was inside. A chitosan-based superabsorbent polymer with monopotassium phosphate was prepared, and it was seen that, within 8 h, all the core fertilizers were released. Like that, potassium nitrate-loaded starch and chitosan-based SAPs were prepared, showing that all the fertilisers were released within 14 days. Low nutrient content and a short duration of release were the major limitations of this method. And the third approach was to coat the fertilizer with several layers that had the capability of water absorption and slow release. However, this approach still had some problems, like a short release period, lower N content, and shedding of the SAP coating. In addition, it caused the risk of environmental pollution if the organic solvents were utilized. Thus, it is important to prepare an environmentally friendly fertiliser with easy processing, high water absorptive ability and N content, and good controlled release performance [20].

Dispat and his co-workers have prepared a bio-based SAP for agricultural applications from modified starch (MS) to increase its biodegradability and antibacterial property. Initially, the starch was modified by tetraethyl orthosilicate and zinc oxide by a sol–gel method under an acidic condition. Then, the modified starch was graft copolymerized with potassium acrylate monomer to prepare a new modified starch-graft-polyacrylate (PA) SAP (MS-g-PA SAP) [12]. Starch has been mostly used to develop the SAPs because of its hydration-swelling property, biodegradability, low cost, and abundance [41]. However, due to the shorter service life of SBSAPs, they become unsuitable for agricultural applications. Thus, starch can be modified by using antibacterial nanoparticles like zinc oxide (ZnO), silver nanoparticles, and titanium dioxide (TiO_2). From these, ZnO is a well-known low-cost material having strong antibacterial and inhibitory properties [92, 93]. However, the dispersion of ZnO nanoparticles in polymers is quite difficult due to their high surface energy and hydrophilic nature, which result in the aggregation and phase separation of the ZnO particles. And interestingly, this problem can be reduced by the silane treatment of the starch surface. Eventually, to get a longer service life and greater antibacterial activity from the SAP, starch was modified by ZnO and tetraethyl orthosilicate (TEOS) by using a sol–gel approach. Then, the modified starch was used to prepare MS-g-PA SAP. The high water holding capacity of the MS-g-PA SAP was known from its high porosity structure, which enables more plants to exist in drought conditions compared to the commercial one [12]. This research confirmed the excellent soil conditioning capacity of MS-g-PA SAP. Zhao and his co-workers have reported the superabsorbent polymers (SAPs) to investigate the effects of SAPs on the agricultural productivity and the water/nutrient holding characteristics of a saline soil. Three types of SAPs were taken for experimentation: a starch-grafted SAP (SGSAP), a commercial SAP (CSAP), and a modified starch-grafted SAP (MSGSAP). SAP application in saline soil improved the nutritional conditions and soil moisture and promoted crop growth. The soil drying-wetting cycle encourages the release and mineralization of the soil nutrients, which enhances nutrient accessibility and availability and promotes root growth [94]. In addition, SAP degradation gives additional exogenous carbon, which has a greater effect on soil carbon and nitrogen mineralisation. Thus, SAP application in the soil boosts soil nutritional conditions and hence promotes crop growth and higher yield [95]. The WAC of SAP is related to its molecular network characteristics. Generally, the higher the cross linking degree of a SAP molecule and the more hydrophilic groups in a SAP, the higher the WAC it has [19]. Moreover, the WAC of SAP is also affected by ion concentration in an absorption medium [96]. Higher concentrations of ions such as Na^+, K^+, Mg^{2+} and Ca^{2+} in the pore water or irrigation water of the salinissed soil would greatly minimise the WAC of SAP. SAPs are generally grafted with natural macromolecules like starch to improve their salt tolerance [97]. From this study, it was assumed that, after being grafted with natural macromolecules, a SA would possess a higher WAC in salinized soils and increase crop production. A salt-tolerant starch-grafted SAP and a commercial SAP were used as soil conditioners in two experiments: a pot test and a soil column leaching experiment [98]. Extensive usage of the superabsorbent polymer (SAP) made from conventional petroleum sources is creating serious threats for the ecosystem. Bora et al. have synthesized a superabsorbent polymer with a high biomaterial content (63%) using itanoic acid and starch as renewable resources. Due to the presence of various polar groups and porous morphology, the synthesized superabsorbent polymer was proven to have a high water absorption capacity. The superabsorbent hydrogels of different compositions were synthesized and their performance towards water absorption in salt solutions and with and without load was examined. The urea-encapsulated superabsorbent hydrogels were employed in soil to determine their capabilities for water capturing capacity, soil porosity enhancement, seed germination rate, etc. Moreover, these superabsorbent hydrogels possessed good biodegradability, which can overcome the limitations of petroleum-based hydrogels. Thus, these have significant potential to be used as environmentally friendly superabsorbent polymers in agricultural applications [31]. Due to the decrease in cultivatable land and increase in the world population, it has become essential for farmers to use large amounts of synthetic fertilizer, mainly urea, potash, nitrogen, superphosphate, potassium, and phosphorus, to enhance crop yield [99]. However, only a small amount of the applied fertiliser is used by the plants, and the rest is quickly leached away, which leads to severe environmental pollution and increases the crop production cost [100]. Hence, researchers have synthesized superabsorbent polymers (SAPs) as a slow-release fertiliser system to overcome such shortcomings in agriculture. The SRF system releases water and nutrients in a controlled manner to the plant, resulting in the enhancement of crop yield as well as reduction of fertiliser loss [101]. Most of the SAPs are prepared from synthetic polymers such as poly (acrylamide) (PAM), poly (acrylic acid) (PAA), or their copolymers. As the synthetic polymers are obtained from non-renewable sources and are not environmentally friendly, focus is given to the preparation of bio-based SAPs to replace the synthetic ones. Bio-based SAPs are considered as superior than the synthetic ones because of the easy availability, biodegradability and biocompatibility attributes of the former [31].

9.2 Application of Starch-Based Superabsorbent Polymer in Road Dust Suppression

Human health and societal progress are both negatively impacted by dust pollution, which is especially problematic in some developing nations. Here, several superabsorbent polymers (SAPs) were prepared using either the traditional liquid phase (LP) or the mechanical activation-assisted solid phase reaction (MASPR) techniques. SAPs synthesized by the two methods were compared with regard to their crystal structure, rheological characteristics, dust suppression performance, and alterations in functional groups. The adsorption experiment between dust particles and dust suppressant has been used to discuss the SAPs dust suppression mechanism. SAP-MA, in comparison to SAP-LP, showed superior suppression effectiveness because of its low molecular weight, high water absorption, and high grafting rate. It is possible that the reduced crystallinity, anti-consolidation, anti-evaporation, and permeability caused by MA in SAP-MA contributed to this enhancement. In addition, the dust particles and the SAPs effectively adsorbing chemicals, lead to a stable consolidation effect [102].

Air pollution is a major problem that affects not only human health but also the health of the planet and its inhabitants. Air pollution is a major hazard to public health and has far-reaching effects on the weather. Increases in automobile traffic and the release of toxic byproducts from a variety of factories are the primary causes of poor air quality (such as cement, coal, steel, minerals, etc.). The quality of the air in cities is greatly impacted by the dust particles created by vehicles [103]. As a result, water spraying, dust screens, and setting up the Cottrell process are the primary means by which traffic dust is mitigated. When it comes to dust, chemical dust suppression is often the most effective way. Accordingly, experts from a variety of nations have undertaken substantial study on a variety of dust suppression methods, such as adhesive dust suppressants [104], compound dust suppressants, and wetting dust suppressants. Some of these, like the wetting dust suppressant, can increase the dust-wetting effect. The issues of severe corrosion, prolonged cure time, and inadequate adhesion persist, though. In addition, corrosive rainwater can easily wash away additives like halides, which can be harmful to transportation infrastructure. Due to their high resistance to evaporation and high quality, bonded chemical dust suppressors containing asphalt, residual oil, and crude oil have been extensively employed though. In addition, corrosive rainwater can easily wash away additives like halides, which can be harmful to transportation infrastructure. Due to their high resistance to evaporation and high quality, bonded chemical dust suppressors containing asphalt, residual oil, and crude oil have been extensively employed. Unfortunately, it is difficult to decay, pollutes soil and water, is poisonous, and has been abandoned due to these characteristics. The composite dust suppressant is effective at wetting, coagulation, and bonding, but its use is severely limited by the absence of optimal preparation procedures [105]. Therefore, there is a pressing need to find a way to control the dust that won't break the bank or harm the ecosystem. Due to its superior moisture-conserving properties, non-toxicity, and viscosity, SAP has recently become the focus of attention for its use in chemical dust suppression [106]. However, the majority of these SAPs continue to resist due to difficulties with insufficient permeability, poor solubility and high surface tension. Moreover, the aqueous solution polymerization, microwave irradiation techniques, and inverse-phase suspension polymerization are the most common synthetic methods employed for the production of SAP. Production on a large scale is typically hindered by the requirement for harsh experimental conditions, a laborious process, a longer reaction time, and a wide variety of devil's liquor [106]. Therefore, in order to increase the yield of SAP dust-suppressor, there is a need for the creation of a cheap and practical processing method that can handle these kinds of technological restrictions.

The solid-phase reaction (SPR) has attracted a lot of attention due to its advantageous characteristics. These characteristics include its simplicity, high efficiency, and independence from the use of a solvent. Heat transfer and poor mass in SPR are still problems that need to be addressed, but primarily for materials with solid structures. Poor uniformity and low efficiency result from the reaction reagent's inability to make precise contact and react with the solid components. Mechanical activation (MA) reduces starch particle size and destroys its crystalline structure, resulting in the creation of highly active –OH groups that increase starch's reactivity. At the same time as it boosts starch modification efficiency, MA improves the contacting condition between the reactants. With the MASPR (mechanical activation-assisted solid phase reaction) technique, the reaction system's materials are directly affected by the vigorous ball-milling. The starch granules are broken down by the strong mechanical functions between the milling balls, which in turn break down the inter- and intramolecular hydrogen bonds, reducing the product's molecular weight and crystallinity. Moreover, the procedure simplifies the operation and makes use of mechanical energy without the need for extra solvents in the reaction. The dust suppression field benefits greatly from the decreased viscosity and better flow-ability of modified starch (MS) [107, 108]. The SBSAP with superior agglomeration and wettability properties was created by Liang et al. using the MASPR procedure; the liquid polymerization (LP) technique was used for comparison. The characterization of the crystalline structure, rheological property, and molecular structure of SAP-MA and SAP-LP produced in different ways allowed us to learn about their distinctive physicochemical properties. These two SAPs were put through their paces in a comprehensive study of permeability, anti-evaporation, and vibration resistance. Evidence suggests that pretreatment of starch with the MASPR procedure can increase the grafting ratio between the starch and PAA compared to the LP approach. When trying to improve hygro-scopicity and moisture-conserving capability, the grafting ratio is a key factor. When subjected to mechanical force, SAP-MA exhibits greater permeability and film-forming characteristics despite its weak gel structure and low molecular weight. Both the quasi-second-order and pseudo-first-order kinetic models provided a linear fit to the data for the adsorption capacities of SAP-LP and SAP-MA, respectively. The quasi-second-order kinetic model well represents the adsorption behavior of the SAP dust suppressor. The equilibrium adsorption capacity of SAP-MA is greater than that of SAP-LP. Large molecule structures easily generate steric hindrance, which decreases adsorption efficiency. The SAP is synthesized using a unique stirring ball mill and the MASPR technique. Compared to SAP made using the traditional liquid phase approach, SAP generated with the MASPR procedure has a lower molecular weight, lower crystallinity, and a higher content of substituted groups. As a result, the SAP dust suppressant's permeability, vibration resistance, and anti-evaporation are all enhanced by the enhancement of its film-forming property and fluidity [102].

10 Conclusions

Superabsorbent polymers (SAPs) are macromolecules that can slowly release water to the environment after absorbing it. There are numerous industrial fields have been interested in superabsorbent polymers (SAPs) because of their potentially beneficial qualities. These extraordinary materials are able to absorb and store vast quantities of water or aqueous solutions without disintegrating. However, the widespread use of petrochemicals in most commercial SAPs on the market today causes significant environmental harm. For the structure of the polymer, we selected to use biodegradable and renewable elements like vegetable starch and fibres derived from cellulose that had been chemically changed. Carbohydrates, which are both plentiful and renewable, are excellent building blocks for the synthesis of biodegradable polymers due to their simplicity. Starch-based SAPs stand out from the other types because of their exceptional mix of qualities, including non-toxicity, biocompatibility, biodegradability, and renewability. In addition to this, there is an abundance of inexpensive starch. In order to turn cassava starch into a semi-synthetic superabsorbent polymer, alkaline saponification and graft copolymerization using an acrylic monomer and a free radical initiator were used as the two primary transformation processes. Therefore, natural alternatives to SAPs, notably polysaccharides, are being studied extensively for their potential environmental benefits. Starch has been shown to be useful in a wide variety of contexts due to its low price, ease of preparation, and widespread availability. In addition, cassava starch has the potential to be polysaccharides-based SAPs due to its high swelling capacity and gel-forming ability. The current reports of SAPs materials, preparation procedures, characteristics, and applications of starch-based super absorbent polymers are presented in this chapter.

Acknowledgements The authors gratefully acknowledge DST-SERB for financial support obtained through project grant of (CRG/2018/004101), New Delhi, Government of India.

Conflict of Interest The authors declare that we have no conflict of interest.

References

1. Zou, W., Yu, L., Liu, X., Chen, L., Zhang, X., Qiao, D., Zhang, R.: Effects of amylose/amylopectin ratio on starch-based superabsorbent polymers. Carbohydr. Polym. **87**(2), 1583–1588 (2012)
2. Peng, M.C., Sethu, V., Selvarajoo, A.: Performance study of chia seeds, chia flour and Mimosa pudica hydrogel as polysaccharide-based superabsorbent polymers for sanitary napkins. Mater. Today Commun. **26**, 101712 (2021)
3. Kabiri, K., Zohuriaan-Mehr, M.: Superabsorbent hydrogel composites. Polym. Adv. Technol. **14**(6), 438–444 (2003)
4. Lee, J.S., Kumar, R.N., Rozman, H.D., Azemi, B.M.N.: Pasting, swelling and solubility properties of UV initiated starch-graft-poly (AA). Food Chem. **91**(2), 203–211 (2005)

5. Mu, Z., Liu, D., Lv, J., Chai, D.F., Bai, L., Zhang, Z., Zhang, W.: Insight into the highly efficient adsorption towards cationic methylene blue dye with a superabsorbent polymer modified by esterified starch. J. Environ. Chem. Eng. **10**(5), 108425 (2022)
6. Qiao, D., Tu, W., Wang, Z., Yu, L., Zhang, B., Bao, X., Lin, Q.: Influence of crosslinker amount on the microstructure and properties of starch-based superabsorbent polymers by one-step preparation at high starch concentration. Int. J. Biol. Macromol. **129**, 679–685 (2019)
7. Zhang, Q., Wang, Z., Zhang, C., Aluko, R.E., Yuan, J., Ju, X., He, R.: Structural and functional characterization of rice starch-based superabsorbent polymer materials. Int. J. Biol. Macromol. **153**, 1291–1298 (2020)
8. Xiao, X., Yu, L., Xie, F., Bao, X., Liu, H., Ji, Z., Chen, L.: One-step method to prepare starch-based superabsorbent polymer for slow release of fertilizer. Chem. Eng. J. **309**, 607–616 (2017)
9. Petroudy, S.R.D., Kahagh, S.A., Vatankhah, E.: Environmentally friendly superabsorbent fibers based on electrospun cellulose nanofibers extracted from wheat straw. Carbohydr. Polym. **251**, 117087 (2021)
10. Li, A., Zhang, J., Wang, A.: Utilization of starch and clay for the preparation of superabsorbent composite. Biores. Technol. **98**(2), 327–332 (2007)
11. Mahmoodi-Babolan, N., Nematollahzadeh, A., Heydari, A., Merikhy, A.: Bioinspired catecholamine/starch composites as superadsorbent for the environmental remediation. Int. J. Biol. Macromol. **125**, 690–699 (2019)
12. Dispat, N., Poompradub, S., Kiatkamjornwong, S.: Synthesis of ZnO/SiO_2-modified starch-graft-polyacrylate superabsorbent polymer for agricultural application. Carbohydr. Polym. **249**, 116862 (2020)
13. Barajas-Ledesma, R.M., Hossain, L., Wong, V.N., Patti, A.F., Garnier, G.: Effect of the counter-ion on nanocellulose hydrogels and their superabsorbent structure and properties. J. Colloid Interface Sci. **599**, 140–148 (2021)
14. González-Henríquez, C.M., Sarabia-Vallejos, M.A., Terraza, C.A., del Campo-García, A., Lopez-Martinez, E., Cortajarena, A.L., Rodríguez-Hernández, J.: Design and fabrication of biocompatible wrinkled hydrogel films with selective antibiofouling properties. Mater. Sci. Eng. C **97**, 803–812 (2019)
15. Fan, S., Tang, Q., Wu, J., Hu, D., Sun, H., Lin, J.: Two-step synthesis of polyacrylamide/poly (vinyl alcohol)/polyacrylamide/graphite interpenetrating network hydrogel and its swelling, conducting and mechanical properties. J. Mater. Sci. **43**, 5898–5904 (2008)
16. Petroudy, S.R.D., Ranjbar, J., Garmaroody, E.R.: Eco-friendly superabsorbent polymers based on carboxymethyl cellulose strengthened by TEMPO-mediated oxidation wheat straw cellulose nanofiber. Carbohydr. Polym. **197**, 565–575 (2018)
17. Bashari, A., Rouhani Shirvan, A., Shakeri, M.: Cellulose-based hydrogels for personal care products. Polym. Adv. Technol. **29**(12), 2853–2867 (2018)
18. Kuang, J., Yuk, K.Y., Huh, K.M.: Polysaccharide-based superporous hydrogels with fast swelling and superabsorbent properties. Carbohydr. Polym. **83**(1), 284–290 (2011)
19. Guilherme, M.R., Aouada, F.A., Fajardo, A.R., Martins, A.F., Paulino, A.T., Davi, M.F., Muniz, E.C.: Superabsorbent hydrogels based on polysaccharides for application in agriculture as soil conditioner and nutrient carrier: a review. Eur. Polymer J. **72**, 365–385 (2015)
20. Zhao, C., Tian, H., Zhang, Q., Liu, Z., Zhang, M., Wang, J.: Preparation of urea-containing starch-castor oil superabsorbent polyurethane coated urea and investigation of controlled nitrogen release. Carbohydr. Polym. **253**, 117240 (2021)
21. Shah, L.A., Khan, M., Javed, R., Sayed, M., Khan, M.S., Khan, A., Ullah, M.: Superabsorbent polymer hydrogels with good thermal and mechanical properties for removal of selected heavy metal ions. J. Clean. Prod. **201**, 78–87 (2018)
22. Khan, H., Chaudhary, J.P., Meena, R.: Anionic carboxymethylagarose-based pH-responsive smart superabsorbent hydrogels for controlled release of anticancer drug. Int. J. Biol. Macromol. **124**, 1220–1229 (2019)

23. Chang, C., Zhang, L.: Cellulose-based hydrogels: present status and application prospects. Carbohydr. Polym. **84**(1), 40–53 (2011)
24. Matalanis, A., Decker, E.A., McClements, D.J.: Inhibition of lipid oxidation by encapsulation of emulsion droplets within hydrogel microspheres. Food Chem. **132**(2), 766–772 (2012)
25. Chang, C., Duan, B., Cai, J., Zhang, L.: Superabsorbent hydrogels based on cellulose for smart swelling and controllable delivery. Eur. Polymer J. **46**(1), 92–100 (2010)
26. Wang, Q., Wilfong, W.C., Kail, B.W., Yu, Y., Gray, M.L.: Novel polyethylenimineacrylamide/ SiO_2 hybrid hydrogel sorbent for rare-earth-element recycling from aqueous sources. ACS Sustain. Chem. Eng. **5**(11), 10947–10958 (2017)
27. Liu, X., Luan, S., Li, W.: Utilization of waste hemicelluloses lye for superabsorbent hydrogel synthesis. Int. J. Biol. Macromol. **132**, 954–962 (2019)
28. Montesano, F.F., Parente, A., Santamaria, P., Sannino, A., Serio, F.: Biodegradable superabsorbent hydrogel increases water retention properties of growing media and plant growth. Agric. Agric. Sci. Procedia **4**, 451–458 (2015)
29. Li, S., Chen, G.: Agricultural waste-derived superabsorbent hydrogels: preparation, performance, and socioeconomic impacts. J. Clean. Prod. **251**, 119669 (2020)
30. Qiao, D., Yu, L., Bao, X., Zhang, B., Jiang, F.: Understanding the microstructure and absorption rate of starch-based superabsorbent polymers prepared under high starch concentration. Carbohydr. Polym. **175**, 141–148 (2017)
31. Bora, A., Karak, N.: Starch and itaconic acid-based superabsorbent hydrogels for agricultural application. Eur. Polymer J. **176**, 111430 (2022)
32. Hao, A., Geng, Y., Xu, Q., Lu, Z., Yu, L.: Study of different effects on foaming process of biodegradable PLA/starch composites in supercritical/compressed carbon dioxide. J. Appl. Polym. Sci. **109**, 2679–2686 (2008)
33. Zhong, K., Lin, Z.T., Zheng, X.L., Jiang, G.B., Fang, Y.S., Mao, X.Y., Liao, Z.W.: Starch derivative-based superabsorbent with integration of water-retaining and controlled-release fertilizers. Carbohydr. Polym. **92**, 1367–1376 (2013)
34. Sotelo-Navarro, P.X., Poggi-Varaldo, H.M., Turpin-Marion, S.J., Rinderknecht Seijas, N.F.: Sodium polyacrylate inhibits fermentative hydrogen production from waste diaper-like material. J. Chem. Technol. Biotechnol. **95**(1), 78–85 (2020)
35. Reshma, G., Reshmi, C.R., Nair, S.V., Menon, D.: Superabsorbent sodium carboxymethyl cellulose membranes based on a new cross-linker combination for female sanitary napkin applications. Carbohydr. Polym. **248**, 116763 (2020)
36. Choi, H., Park, J., Lee, J.: Sustainable bio-based superabsorbent polymer: poly (itaconic acid) with superior swelling properties. ACS Appl. Polym. Mater. **4**(6), 4098–4108 (2022)
37. Ali, S.S., Tang, X., Alavi, S., Faubion, J.: Structure and physical properties of starch/poly vinyl alcohol/sodium montmorillonite nanocomposite films. J. Agric. Food Chem. **59**(23), 12384–12395 (2011)
38. Hu, X.P., Zhang, B., Jin, Z.Y., Xu, X.M., Chen, H.Q.: Effect of high hydrostatic pressure and retrogradation treatments on structural and physicochemical properties of waxy wheat starch. Food Chem. **232**, 560–565 (2017)
39. Sarkar, A., Biswas, D.R., Datta, S.C., Dwivedi, B.S., Bhattacharyya, R., Kumar, R., Patra, A.K.: Preparation of novel biodegradable starch/poly (vinyl alcohol)/bentonite grafted polymeric films for fertilizer encapsulation. Carbohydr. Polym. **259**, 117679 (2021)
40. Taggart, P., Mitchell, J.R.: Starch. In: Handbook of Hydrocolloids, pp. 108–141 (2009)
41. Dave, P.N., Gor, A.: Natural polysaccharide-based hydrogels and nanomaterials: recent trends and their applications. In: Handbook of Nanomaterials for Industrial Applications, pp. 36–66 (2018)
42. Güçlü, G., Al, E., Emik, S., İyim, T.B., Özgümüş, S., Özyürek, M.: Removal of Cu^{2+} and Pb^{2+} ions from aqueous solutions by starch-graft-acrylic acid/montmorillonite superabsorbent nanocomposite hydrogels. Polym. Bull. **65**(4), 333–346 (2010)
43. Parvathy, P.C., Jyothi, A.N.: Synthesis, characterization and swelling behaviour of superabsorbent polymers from cassava starch-graft-poly (acrylamide). Starch-Stärke **64**(3), 207–218 (2012)

44. Alharbi, K., Ghoneim, A., Ebid, A., El-Hamshary, H., El-Newehy, M.H.: Controlled release of phosphorous fertilizer bound to carboxymethyl starch-g-polyacrylamide and maintaining a hydration level for the plant. Int. J. Biol. Macromol. **116**, 224–231 (2018)
45. El-saied, H.A.A., El-Fawal, E.M.: Green superabsorbent nanocomposite hydrogels for high-efficiency adsorption and photo-degradation/reduction of toxic pollutants from waste water. Polym. Test. **97**, 107134 (2021)
46. Qiao, D., Zou, W., Liu, X., Yu, L., Chen, L., Liu, H., Zhang, N.: Starch modification using a twin-roll mixer as a reactor. Starch-Stärke **64**, 821–825 (2012)
47. Moad, G.: Chemical modification of starch by reactive extrusion. Prog. Polym. Sci. **36**, 218–237 (2011)
48. Olivato, J.B., Grossmann, M.V.E., Yamashita, F., Eiras, D., Pessan, L.A.: Citric acid and maleic anhydride as compatibilizers in starch/poly(butylene adipate-coterephthalate) blends by one-step reactive extrusion. Carbohydr. Polym. **87**, 2614–2618 (2012)
49. Hablot, E., Dewasthale, S., Zhao, Y., Zhiguan, Y., Shi, X., Graiver, D., Narayan, R.: Reactive extrusion of glycerylated starch and starch–polyester graft copolymers. Eur. Polymer J. **49**, 873–881 (2013)
50. Qureshi, M.A., Nishat, N., Jadoun, S., Ansari, M.Z.: Polysaccharide based superabsorbent hydrogels and their methods of synthesis: a review. Carbohydr. Polym. Technol. Appl. **1**, 100014 (2020)
51. Kabiri, K., Omidian, H., Hashemi, S.A., Zohuriaan-Mehr, M.J.: Synthesis of fast-swelling superabsorbent hydrogels: effect of crosslinker type and concentration on porosity and absorption rate. Eur. Polymer J. **39**, 1341–1348 (2003)
52. Rubinstein, M., Panyukov, S.: Elasticity of polymer networks. Macromolecules **35**, 6670–6686 (2002)
53. Lanthong, P., Nuisin, R., Kiatkamjornwong, S.: Graft copolymerization, characterization, and degradation of cassava starch-g-acrylamide/itaconic acid superabsorbents. Carbohydr. Polym. **66**(2), 229–245 (2006)
54. Athawale, V.D., Lele, V.: Factors influencing absorbent properties of saponified starch-g-(acrylic acid-co-acrylamide). J. Appl. Polym. Sci. **77**(11), 2480–2485 (2000)
55. Çaykara, T., Bozkaya, U., Kantoğlu, Ö.: Network structure and swelling behavior of poly (acrylamide/crotonic acid) hydrogels in aqueous salt solutions. J. Polym. Sci. Part B: Polym. Phys. **41**(14), 1656–1664 (2003)
56. Kiatkamjornwong, S., Mongkolsawat, K., Sonsuk, M.: Synthesis and property characterization of cassava starch grafted poly [acrylamide-co-(maleic acid)] superabsorbent via γ-irradiation. Polymer **43**(14), 3915–3924 (2002)
57. Sen, M., Yakar, A., Güven, O.: Determination of average molecular weight between cross-links (Mc) from swelling behaviours of diprotic acid-containing hydrogels. Polymer **40**(11), 2969–2974 (1999)
58. Athawale, V.D., Lele, V.: Graft copolymerization onto starch. II. Grafting of acrylic acid and preparation of it's hydrogels. Carbohydr. Polym. **35**(1–2), 21–27 (1998)
59. Karadağ, E., Üzüm, Ö.B., Saraydin, D.: Water uptake in chemically crosslinked poly (acrylamide-co-crotonic acid) hydrogels. Mater. Des. **26**(4), 265–270 (2005)
60. Mostafa, K.M.: Graft polymerization of acrylic acid onto starch using potassium permanganate acid (redox system). J. Appl. Polym. Sci. **56**(2), 263–269 (1995)
61. Athawale, V.D., Lele, V.: Syntheses and characterisation of graft copolymers of maize starch and methacrylonitrile. Carbohydr. Polym. **41**(4), 407–416 (2000)
62. Fanta, G.F.: Synthesis of graft and block copolymers of starch. Block Graft Copolym. **1**(11) (1973)
63. Rather, R.A., Bhat, M.A., Shalla, A.H.: An insight into synthetic, physiological aspect of superabsorbent hydrogels based on carbohydrate type polymers for various applications: a review. Carbohydr. Polym. Technol. Appl. 100202 (2022)
64. Ma, Y.C., Manolache, S., Sarmadi, M., Denes, F.S.: Synthesis of starch copolymers by silicon tetrachloride plasma-induced graft polymerization. Starch-Stärke **56**(2), 47–57 (2004)

65. Shaikh, M.M., Lonikar, S.V.: Starch–acrylics graft copolymers and blends: synthesis, characterization, and applications as matrix for drug delivery. J. Appl. Polym. Sci. **114**(5), 2893–2900 (2009)
66. Nakason, C., Wohmang, T., Kaesaman, A., Kiatkamjornwong, S.: Preparation of cassava starch-graft-polyacrylamide superabsorbents and associated composites by reactive blending. Carbohydr. Polym. **81**(2), 348–357 (2010)
67. Kaith, B.S., Jindal, R., Jana, A.K., Maiti, M.: Development of corn starch based green composites reinforced with Saccharum spontaneum L fiber and graft copolymers—evaluation of thermal, physico-chemical and mechanical properties. Biores. Technol. **101**(17), 6843–6851 (2010)
68. Meshram, M.W., Patil, V.V., Mhaske, S.T., Thorat, B.N.: Graft copolymers of starch and its application in textiles. Carbohydr. Polym. **75**(1), 71–78 (2009)
69. Mostafa, K.H.M., Morsy, M.S.: Modification of carbohydrate polymers via grafting of methacrylonitrile onto pregelled starch using potassium monopersulfate/Fe^{2+} redox pair. Polym. Int. **53**(7), 885–890 (2004)
70. Salam, A., Pawlak, J.J., Venditti, R.A., El-tahlawy, K.: Synthesis and characterization of starch citrate−chitosan foam with superior water and saline absorbance properties. Biomacromolecules **11**(6), 1453–1459 (2010)
71. Comer, C. M., Jessop, J. L.: Evaluation of Novel Back-flush Filtration for Removal of Homopolymer from Starch-g-PMMA. Starch-Stärke, **60**(7), 335–339 (2008)
72. Rath, S. K., Singh, R. P.: Flocculation characteristics of grafted and ungrafted starch, amylose, and amylopectin. J. Appl. Polym. Sci., **66**, 1721–1729 (1997)
73. Della Valle, G., Vergnes, B., Lourdin, D.: Viscous properties of thermoplastic starches from different botanical origin. Int. Polym. Proc. **22**(5), 471–479 (2007)
74. Liu, H., Yu, L., Simon, G., Dean, K., Chen, L.: Effects of annealing on gelatinization and microstructures of corn starches with different amylose/amylopectin ratios. Carbohydr. Polym. **77**(3), 662–669 (2009)
75. Freitas, R.A., Paula, R.C., Feitosa, J.P.A., Rocha, S., Sierakowski, M.R.: Amylose contents, rheological properties and gelatinization kinetics of yam (Dioscorea alata) and cassava (Manihot utilissima) starches. Carbohydr. Polym. **55**(1), 3–8 (2004)
76. Ni, B., Liu, M., Lu, S., Xie, L., Wang, Y.: Environmentally friendly slow-release nitrogen fertilizer. J. Agric. Food Chem. **59**(18), 10169–10175 (2011)
77. Piao, S., Ciais, P., Huang, Y., Shen, Z., Peng, S., Li, J., Fang, J.: The impacts of climate change on water resources and agriculture in China. Nature **467**(7311), 43–51 (2010)
78. Marcì, G., Mele, G., Palmisano, L., Pulito, P., Sannino, A.: Environmentally sustainable production of cellulose-based superabsorbent hydrogels. Green Chem. **8**(5), 439–444 (2006)
79. Nykänen, V.P.S., Nykänen, A., Puska, M.A., Silva, G.G., Ruokolainen, J.: Dual-responsive and super absorbing thermally cross-linked hydrogel based on methacrylate substituted polyphosphazene. Soft Matter. **7**(9), 4414–4424 (2011)
80. Omidian, H., Rocca, J.G., Park, K.: Advances in superporous hydrogels. J. Control. Release **102**(1), 3–12 (2005)
81. Zhong, K., Lin, Z.T., Zheng, X.L., Jiang, G.B., Fang, Y.S., Mao, X.Y., Liao, Z.W.: Starch derivative-based superabsorbent with integration of water-retaining and controlled-release fertilizers. Carbohydr. Polym. **92**(2), 1367–1376 (2013)
82. Ni, B., Liu, M., Lü, S.: Multifunctional slow-release urea fertilizer from ethylcellulose and superabsorbent coated formulations. Chem. Eng. J. **155**, 892–898 (2009)
83. Chang, L., Xu, L., Liu, Y., Qiu, D.: Superabsorbent polymers used for agricultural water retention. Polym. Test. **94**, 107021 (2021)
84. Fernández-Escobar, R., Benlloch, M., Herrera, E., Garcıa-Novelo, J.M.: Effect of traditional and slow-release N fertilizers on growth of olive nursery plants and N losses by leaching. Sci. Hortic. **101**, 39–49 (2004)
85. Naz, M.Y., Sulaiman, S.A.: Slow release coating remedy for nitrogen loss from conventional urea: a review. J. Control. Release **225**, 109–120 (2016)

86. Yang, Y., Zhang, M., Li, Y.C., Fan, X., Geng, Y.: Controlled release urea improved nitrogen use efficiency, activities of leaf enzymes, and rice yield. Soil Sci. Soc. Am. J. **76**, 2307 (2012)
87. Mohd Ibrahim, K.R., Eghbali Babadi, F., Yunus, R.: Comparative performance of different urea coating materials for slow release. Particuology **17**, 165–172 (2014)
88. Sartore, L., Vox, G., Schettini, E.: Preparation and performance of novel biodegradable polymeric materials based on hydrolyzed proteins for agricultural application. J. Polym. Environ. **21**, 718–725 (2013)
89. Rashidzadeh, A., Olad, A.: Slow-released NPK fertilizer encapsulated by NaAlg-g-poly (AA-co-AAm)/MMT superabsorbent nanocomposite. Carbohydr. Polym. **114**, 269–278 (2014)
90. Yasmin, N., Blair, G., Till, R.: Effect of elemental sulfur, gypsum, and elemental sulfur coated fertilizers, on the availability of sulfur to rice. J. Plant Nutr. **30**(1), 79–91 (2007)
91. Sempeho, S.I., Kim, H.T., Mubofu, E., Hilonga, A.: Meticulous overview on the controlled release fertilizers (2014)
92. Pullagurala, V.L.R., Adisa, I.O., Rawat, S., Kim, B., Barrios, A.C., Medina-Velo, I.A., Gardea-Torresdey, J.L.: Finding the conditions for the beneficial use of ZnO nanoparticles towards plants—a review. Environ. Pollut. **241**, 1175–1181 (2018)
93. Abbatt, J.P., Leaitch, W.R., Aliabadi, A.A., Bertram, A.K., Blanchet, J.P., Boivin-Rioux, A., Yakobi-Hancock, J.D.: Overview paper: new insights into aerosol and climate in the Arctic. Atmos. Chem. Phys. **19**(4), 2527–2560 (2019)
94. Zia, R., Nawaz, M.S., Siddique, M.J., Hakim, S., Imran, A.: Plant survival under drought stress: implications, adaptive responses, and integrated rhizosphere management strategy for stress mitigation. Microbiol. Res. **242**, 126626 (2021)
95. Satriani, A., Catalano, M., Scalcione, E.: The role of superabsorbent hydrogel in bean crop cultivation under deficit irrigation conditions: a case-study in Southern Italy. Agric. Water Manag. **195**, 114–119 (2018)
96. Thombare, N., Mishra, S., Siddiqui, M.Z., Jha, U., Singh, D., Mahajan, G.R.: Design and development of guar gum based novel, superabsorbent and moisture retaining hydrogels for agricultural applications. Carbohydr. Polym. **185**, 169–178 (2018)
97. Irshad, M.A., Nawaz, R., ur Rehman, M.Z., Adrees, M., Rizwan, M., Ali, S., Tasleem, S.: Synthesis, characterization and advanced sustainable applications of titanium dioxide nanoparticles: a review. Ecotoxicol. Environ. Saf. **212**, 111978 (2021)
98. Zhao, C., Zhang, L., Zhang, Q., Wang, J., Wang, S., Zhang, M., Liu, Z.: The effects of bio-based superabsorbent polymers on the water/nutrient retention characteristics and agricultural productivity of a saline soil from the Yellow River Basin, China. Agric. Water Manag. **261**, 107388 (2022)
99. Wen, P., Wu, Z., Han, Y., Cravotto, G., Wang, J., Ye, B.C.: Microwave-assisted synthesis of a novel biochar-based slow-release nitrogen fertilizer with enhanced waterretention capacity. ACS Sustain. Chem. Eng. **5**(8), 7374–7382 (2017)
100. Dhanapal, V., Subhapriya, P., Nithyanandam, K.P., Kiruthika, M.V., Keerthana, T., Dineshkumar, G.: Design, synthesis and evaluation of N,N1-methylenebisacrylamide cross-linked smart polymer hydrogel for the controlled release of water and plant nutrients in agriculture field. Mater. Today: Proc. **45**, 2491–2497 (2021)
101. Hakim, S., Darounkola, M.R.R., Barghemadi, M., Parvazinia, M.: Fabrication of PVA/nanoclay hydrogel nanocomposites and their microstructural effect on the release behavior of a potassium phosphate fertilizer. J. Polym. Environ. **27**(12), 2925–2932 (2019)
102. Liang, Z., Cai, X., Hu, H., Zhang, Y., Chen, Y., Huang, Z.: Synthesis of starch-based super absorbent polymer with high agglomeration and wettability for applying in road dust suppression. Int. J. Biol. Macromol. **183**, 982–991 (2021)
103. Koh, B., Kim, E.: Comparative analysis of urban road dust compositions in relation to their potential human health impacts. Environ. Pollut. **255**, 113156 (2019)
104. Zhou, G., Ding, J., Ma, Y., Li, S., Zhang, M.: Synthesis and performance characterization of a novel wetting cementing agent for dust control during conveyor transport in coal mines. Powder Technol. **360**, 165–176 (2020)

105. Jin, H., Nie, W., Zhang, H., Liu, Y., Bao, Q., Wang, H., Huang, D.: Preparation and characterization of a novel environmentally friendly coal dust suppressant. J. Appl. Polym. Sci. **136**, 47354 (2019)
106. Bao, Q., Nie, W., Liu, C., Liu, Y., Zhang, H., Wang, H., Jin, H.: Preparation and characterization of a binary-graft-based, water-absorbing dust suppressant for coal transportation. J. Appl. Polym. Sci. **136**, 47065 (2018)
107. Lee, T., Park, J., Knoff, D.S., Kim, K., Kim, M.: Liquid amphiphilic polymer for effective airborne dust suppression. RSC Adv. **9**, 40146–40151 (2019)
108. Mingyue, W., Xiangming, H., Qian, Z., Wei, L., Yanyun, Z., Zhenglong, H.: Study on preparation and properties of environmentally-friendly dust suppressant with semiinterpenetrating network structure. J. Clean. Prod. **259**, 120870 (2020)

Smart Superabsorbents and Other Bio-based Superabsorbents

Shubhasmita Rout

Abstract Superabsorbent polymers are 3-dimensional hydrophilic polymeric materials with the capacity to hold large amounts of fluids i.e. can be 1000 times their original mass by maintaining their dimensional stability. These water-absorbing hydrogels when mixed with an aqueous solution absorb and retain fluids by forming hydrogen bonds with the water molecules. The absorbance and swelling capacity can be modified by changing the concentration of crosslinking polymers. To meet the desirable property for various applications, a considerable number of studies are conducted to develop some polymers with special chemical properties termed "Smart superabsorbents". These are the class of advanced polymers with the ability to change their physical properties in response to environmental stimuli such as a change in Temperature, Pressure, pH, Light, etc. Smart superabsorbent hydrogels have tremendous applications in bio-medical, water purification, forming actuators, and conductive hydrogels due to their fast macroscopic changes in structure and reversible transition properties. These are synthetic polymers derived from petroleum products such as acrylates and acrylamides with high production costs and toxicity. To minimize the toxicity researchers are moving their interest towards bio-friendly hydrogels derived from living organisms or chemically produced from biological materials. There are different polysaccharides-based SAPs such as alginate, cellulose, chitosan, etc. used to form hydrogels, which pose less water absorption capacity and poor life span in comparison with the synthetic one. So the focus is shifted towards making semi-synthetic SAP to target the interest of specialized applications where external stimuli are used to create fine-tuned SAP properties. This chapter covers approaches adopted from the literature to study semi-synthetic smart superabsorbents with minimal toxicity.

The original version of this chapter was revised: The affiliation of the author Shubhasmita Rout has been corrected. The correction to this chapter is available at
https://doi.org/10.1007/978-981-99-3094-4_11

S. Rout (✉)
CIPET-IPT, Bhubaneswar, Odisha, India
e-mail: subhasmita.rout@che.iith.ac.in

S. Pradhan and S. Mohanty (eds.), *Bio-based Superabsorbents*, Engineering Materials,
https://doi.org/10.1007/978-981-99-3094-4_8

Keywords Smart superabsorbents · Hydrogel · Swellability · Toxicity · Bio-based superabsorbents

1 Introduction

Superabsorbent Polymers (SAPs) have gained popularity due to their potential applications in baby diapers, sanitary products, adult diapers, etc. [1]. The advanced version of SAPs termed "Smart Superabsorbent Polymer" has brought the attention of researchers due to its extraordinary flexibility to change properties by sensing the external stimuli, making this a suitable material for various applications including in ice packs, food packaging materials, fertilizer coating agents, agricultural and horticultural fields (soil repair and desert greening agents), Self-healing and self-sealing Concretes, Bio-sensors, waste-water treatment, dye-removal and so on [2, 3]. Due to the differences in the degree of swelling in response to changes in pH, SAPs have been widely applied in various biomedical applications, such as in pharmaceuticals, treatment of edema, drug delivery systems, and expansion microscopy [4–7]. Based on the hygroscopic properties of SAPs, superabsorbent fibers can absorb moisture from the air and have been applied to produce disposable medical clothing materials for the removal of excess moisture during COVID-19 [8]. The network in SAPs is formed by crosslinking polymer chains which are bonded by covalent bonds, hydrogen bonds, or Vander Waals interactions or physical entanglements [9]. The colloidal polymeric solution of SAP can be lyophilic, lyophobic, or amphiphilic depending upon the interaction with solvents. When the polymer molecule contains both the hydrophilic and hydrophobic parts, the amphiphilic nature prevails. Micelles, dendrimers, and star polymers are examples of the polymer chain containing a hydrophilic head and a hydrophobic tail. Again properties of the polymer in SAPs can be altered according to the required characteristics, by grafting with another material. These SAPs are used for the synthesis of hydrogels, aerogels, and microgels (soft colloids) which have various biomedical and agricultural applications. Smart SAP undergoes a large-reversible change by means of a change in physical or chemical properties in response to applied stimuli [10]. The phenomena of smart SAPs involve structural and volume phase transition in response to external stimuli including temperature, pH, ionic concentration, light, magnetic fields, electrical fields, and chemicals. The focus was mostly inclined toward volume phase transitions as it triggers the functions of movements such as deformation, volume change modulus changes, and force generation. These SAPs can respond to a single stimulus or multiple stimuli such as temperature, pH, electric or magnetic field, light intensity, biological molecules, etc., by changing the physical characteristics that show a transition from the swollen (solution) state to collapsed (gel) state [11]. The superabsorbent polymer shows an adaptable and solvent-selective degree of swelling. Natural responsive and adaptive processes are also present in living bodies such as focusing the eye, opening and closing pores, and healing wounds [12]. Most of these responsive and natural processes are directed either by conformational changes or by the aggregation of proteins, referred to as nature's smart polymer. Likewise, synthetic polymers produced from petroleum-based products are the building blocks for the formation of artificial smart polymeric materials. A variety of smart polymers have been developed till now that change their properties by responding to natural stimuli and certain molecules including CO_2 and sugars [13]. The responses shown by these polymers to external stimuli can also appear as a variety of physical properties such as a change in shape, color, or solubility. Thermo-responsive polymers have a wide range of applications in the bio-medical field as variations of temperature that can be applied externally. In nature also throughout the day, spontaneous fluctuations in temperature take place. Similarly, in the case of inflated tissue, temperature also increases [14]. Bio-responsive physicochemical nature of Smart polymers over power is due to conventionally available polymers in medicine due to the controlled release of specific materials. In the last two decades, there are numerous biomedical applications where smart SAPs have been used, including diagnostics, drug delivery, bio-imaging, bio-sensing, cell culture, and tissue engineering (Figs. 1 and 2).

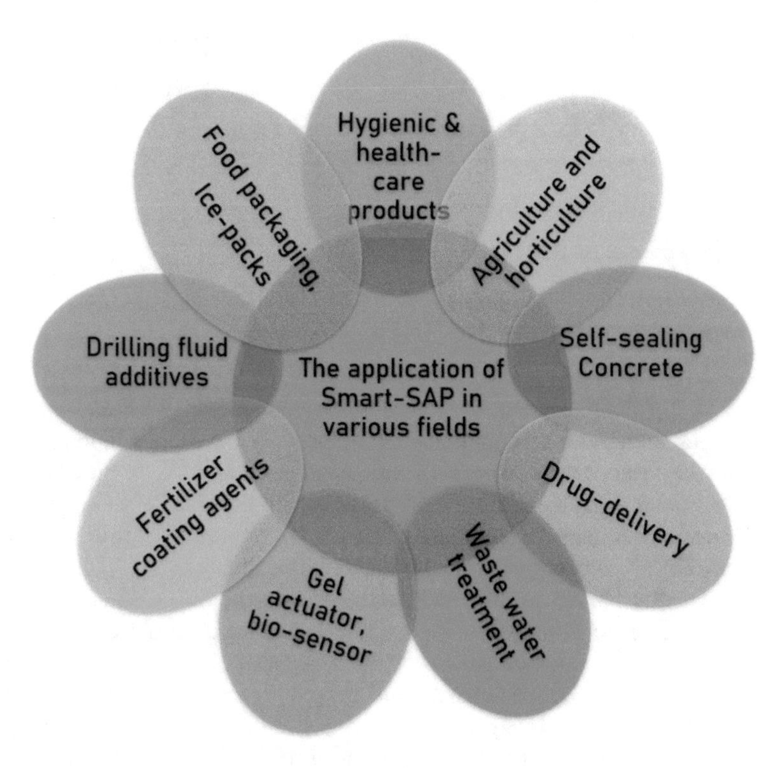

Fig. 1 Applications of smart-SAP

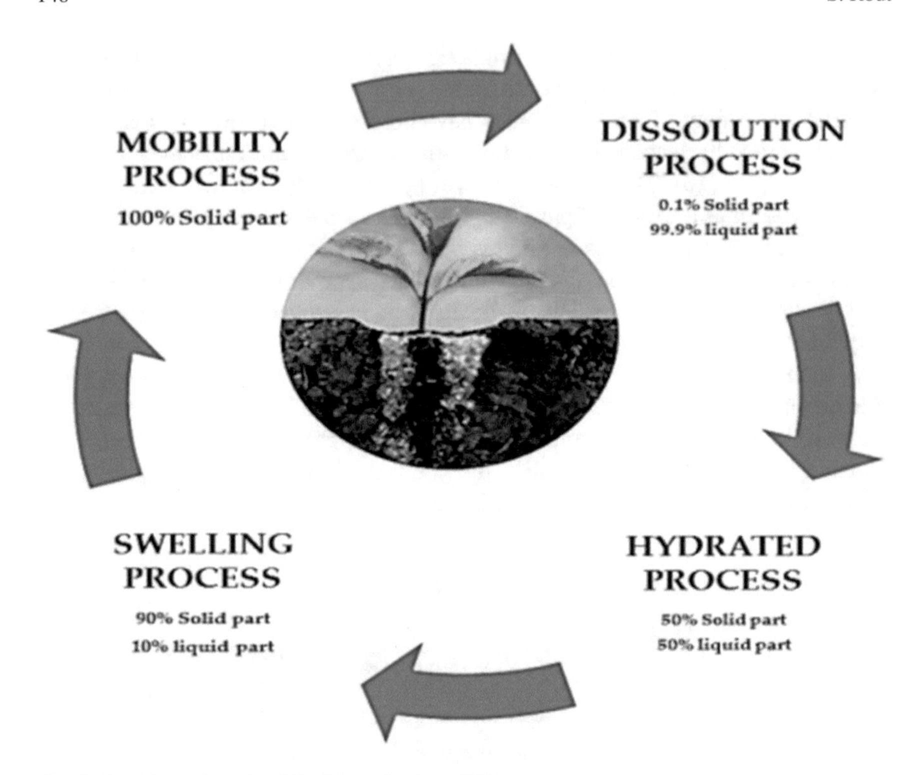

Fig. 2 Working principle of SAP in agriculture [15]

2 Temperature-Responsive SAPs

These are one of the most frequently used smart SAP, in which the aqueous polymeric solution undergoes a phase transition induced by the change in temperature. Along with the phase, structural properties also change. This is the commonly studied responsive system with potential applications in the bio-medical field. Depending on the Critical solution temperature, these polymers are either Lower critical solution temperature (LCST) and Upper critical solution temperature (UCST) [16, 17]. Polymeric components that are miscible and swell in solution below LCST are referred to as negative/inverse temperature sensitivity [18]. If swelling takes place above UCST, it is regarded as positive temperature sensitivity. A case of increased swelling upon an increase in temperature was noticed in an interpenetrating network of poly(acrylic acid) and poly(acrylamide) [19]. The swelling/deswelling action is due to the presence of amphipathic groups (of the polymer itself or developed by grafting) whose hydrophilic segment interacts with water molecules at certain temperature ranges. In the case of LCST, the polymer absorbs the surrounding liquid and starts to swell below the critical temperature and as the temperature increases, the polymer starts to shrink forcing the absorbed liquid out. Some thermo-responsive natural polymer-based SAPs are chitosan, cellulose, Pectin, xyloglucan, poly(N-vinyl caprolactam),

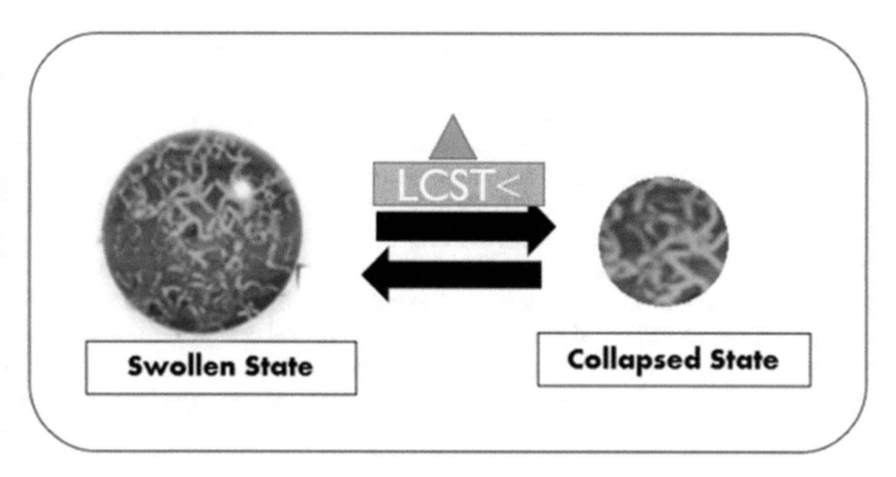

Fig. 3 Swelling/de-swelling behaviour in temperature-responsive SAPs

etc. [20]. For oral drug delivery a pectin-based SAP and for neural tissue engineering and skin regeneration chitosan have already been used [21, 22]. Some examples of petroleum-based SAPs are poly(N-isopropyl acrylamide), poly(2-oxazoline), and poly(N,N-diethyl acrylamide) [23]. To achieve bio-degradability along with low cost, natural polymer-grafted SAPs with thermos-responsive properties are gaining attention in various bio-medical applications (Fig. 3).

3 pH-Responsive SAPs

These are the polymeric materials that change their dimension either by increasing in size (swell) or collapsing in response to a change in the pH value of the surrounding medium. These can be either acidic or basic depending on the type of functional group present in the polymer chain. Depending on the pH value, two types of pH-sensitive materials exist. One has an acidic group (–COOH, –SO3H) of weak polyacids (poly(acrylic acids) or poly(methacrylic acids)) and swells in basic pH, and the other one with a basic group (–NH2) of polybases (poly(N-dimethylaminoethyl methacrylate), poly(N-diethylaminoethyl methacrylate), poly(ethyl pyrrolidine methacrylate)) and swells in acidic pH condition. So the mechanism of responses for both the polymeric solution is same while only the stimulus varies [24, 25]. The basic mechanism involves the moieties in the structure of the polymer in the solution which protonate or deprotonate as a function of pH of that solution. The pendant groups present in the aqueous solution possessing appropriate pH and ionic strength ionize and form fixed charges on the polymer network. This leads to the generation of electrostatic repulsive forces causing the swell or shrinkage of pH-responsive SAPs. Higher swelling capacity is observed in the case of identical charges as repulsive force creates more free space to hold a larger amount of water. Thus pH plays an important role in the controlled release of drugs as a minute change in pH can affect the mesh size of the polymeric network significantly. pH-responsive SAPs, synthesized by the grafting of pH-responsive natural polymers with various synthetic polymers have widespread specialized uses in the drug delivery system, gene carriers or glucose sensors, etc. Starch-poly(sodium acrylate-co-acrylamide) and acryloyl ester of 5-[4-(hydroxy phenyl) azo] salicylic acid (HPAS) have been used in drug delivery systems. Polyacrylic acid-based nanoparticles as a drug delivery agent can be used for cancer therapy [26]. A polysaccharide-based superabsorbent hydrogel isolated from psyllium husk (dietary fiber) showed good swelling capacity in water at different physiological pH values and could be a valuable pharmaceutical excipient for intelligent and targeted drug delivery [27]. Another pH-responsive Superabsorbent polymeric hydrogel based on collagen shows controlled delivery of ephedrine mainly used for bronchial asthma, allergic illnesses, as an antiedemic for mucous membranes in rhinitis, and also as a drug to increase blood pressure during surgical interventions. It is used locally in ophthalmology as a vasoconstricting agent for dilating pupils [28]. Another example of pH-responsive SAP to stimulate the autogenous healing process in mortar poly(acrylic acid-co-acrylamide SAPs, methacrylated alginate, and acid monomers or methacrylated polysaccharides with amine-based monomers are used [29] (Fig. 4).

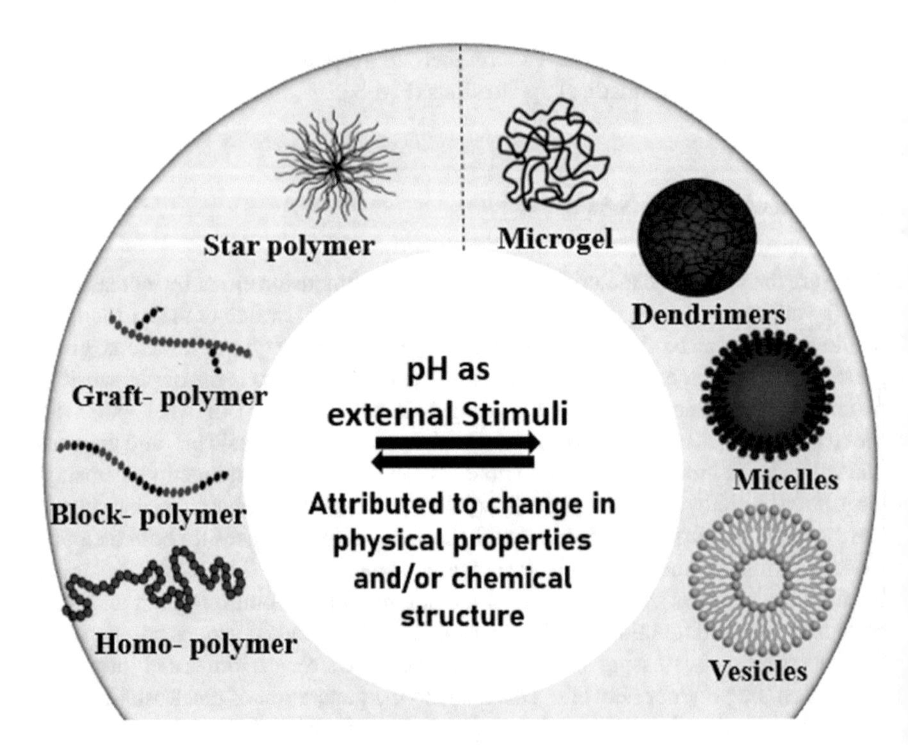

Fig. 4 Phase transition in pH-responsive SAPs

4 Electric Field Responsive SAPs

These are the class of smart SAPs where the polyelectrolyte material changes its structural property (Swell/de-swell) in response to electric current induced by an electric field. These SAPs also show swelling action on one side and shrinkage on the other side of the solution. The degree of swelling completely depends upon the charges of ions and electrodes [30]. The presence of a large number of ionizable groups on the backbone chain made this polymer sensitive to pH and electricity. Depending on the electrical charge present on the backbone and or side chain SAPs are classified into four types [31].

1. Non-ionic (No charge present such as agarose)
2. Ionic (Possess either anionic or cationic moieties. For example acrylates or alginate)
3. Ampholytic (Both acidic and basic functionalities are present)
4. Zwitter-ionic (Containing both anionic and cationic groups. The overall net charge is zero [32].

Polymers having electro-sensitivity convert the chemical/electrical energy into mechanical energy including expansion, contraction, elongation and bending of the SAPs [33–35]. These changes are due to the attraction of ionic groups to the oppositely charged electrodes, induced by an electrical or chemical potential. The applied electric field varies from the solution where SAP is in contact with the electrode to the solution where SAP is placed in water or acetone or their mixture solution even in the presence of electrolytes. Examples of electro-responsive SAPs are sodium alginate-grafted-poly(acrylic acid) poly(acrylamide), poly(ethyloxazoline), polythiophene [36]. These SAPs have potential applications in controlled drug release.

5 Magneto-responsive SAP

The interest in smart SAPs are expanding towards newly emerging triggers such as light, electric and magnetic field from the available ones (temperature, pH) which are limited to mass transfer. The newly developed responsive material comes with more precise responsivity with higher efficiency. Materials including gels, polymers and liquid crystals contain different types of magnetic particles varying in their size that respond to the magnetic field as an external stimulus. Magnetic properties of a material are based on its size, shape, structure, chemistry and crystallinity. The components including iron oxide (Fe_3O_4, Fe_2O_3), pure metals (ferrum, cobalt, nickel) and their alloys and chemical compounds exhibit magnetic response towards the magnetic field. Magnetite (Fe_3O_4) is one of the excellent magnetic nanocomponents with bio-compatibility and nontoxicity having wide range of applications in Ferro fluid, catalysis, pigment color, magnetic storage media, environment protection, cell separation, clinic diagnosis and therapy. Hematite (Fe_2O_3) has application in gas sensors, magnetic recording media due to the transitional property from antiferromagnetic to weak ferromagnetic above the Morin temperature (260 K). Maghemtite (gamma-Fe_2O_3) is a ferrimagnetic in which oxygen atoms form a FCC structure making it an ideal candidate for fabrication of luminescent and magnetic dual functional nanocomposites due to its excellent transparent properties. These are bio-compatible and have applications in cell separation and a drug delivery agent in cancer therapy. Due to their non-toxicity and natural existance in many tissues, Fe3+ ions are the most desired in the formation of intracellular and macromolecular biologically active or magnetic resonance contrast materials. Again magneto-sensitive SAP can realize temporal and remote control having its potential use in microfluidic systems in analytical and therapeutic devices. For in-vivo application, magnetic nanoparticles need to be non-toxic, biocompatible and they should disperse-well in the carrier solvent exhibiting high magnetic saturation. Supermagnetic particles have great interest in in vivo application as they do not retain any residual magnetism after the removal of the magnetic field. In the bio-medical field magnetic nanoparticles are used for drug delivery and magnetic resonance imaging [24].

6 Photo-Responsive SAPs

Photo or Light responsive SAPs are interesting one as they can influence the solubility conformation, polarity, amphiphilicity, charge, optical chirality, and conjugation of the polymer in response to light. The major advantage here is that light stimulus can be delivered directly via laser with high accuracy and ease of control without disturbing the surroundings due to which photo-responsive SAPs have been used in high potential applications such as smart devices, actuators (micro-scale or nano-scale), optical sensors, controlled release of drugs, and separators. This polymer undergoes reversible or irreversible change in conformation. The dissociation of the molecule upon irradiation of UV- light leads to increase in water affinity and electrostatic repulsion of ionic groups thus increasing the swelling action. Similarly when the visible light sensitive chromophore is incorporated it absorbs light and disperses the energy into heat due to the radiation-less transition which causes the shrinkage of SAP by increasing its temperature. When the photo-sensitive chromophore is incorporated to the polymer system there is a transition from microscopic deformation of the chromophore to macroscopic deformation of the polymer induced by photo contractions which convert the light energy to mechanical energy [37]. Some reversible photo-responsive SAPs include Functionalized azobenzene compounds, leuco derivatives, cinnamic acids, and spirobenzopyran derivatives. Azobenzenes can be used as comonomers (side chain or pendant group) or as crosslinkers in liquid crystal networks of polymers and also in interpenetrating or semi-IPN systems due to their reversible photoisomerization property. The phenomena involved here is due to the photochemical stimuli that the more stable trans isomer converts to the less stable cis isomer which can be isomerized by photochemical or thermal stimuli. It is converted to the trans state again by isomerizing, and the latter may subsequently isomerize back to the trans-state either thermally or photochemically [38] (Figs. 5 and 6).

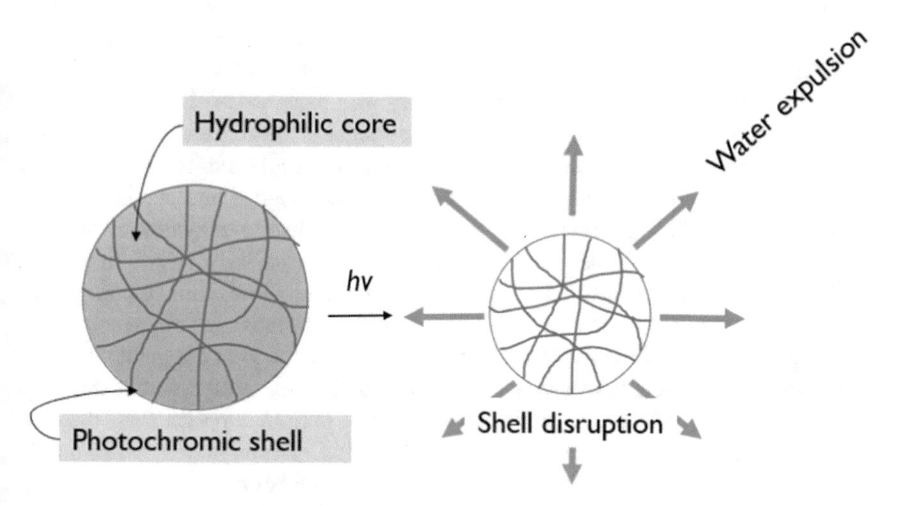

Fig. 5 Effect of light on photo-responsive SAPs

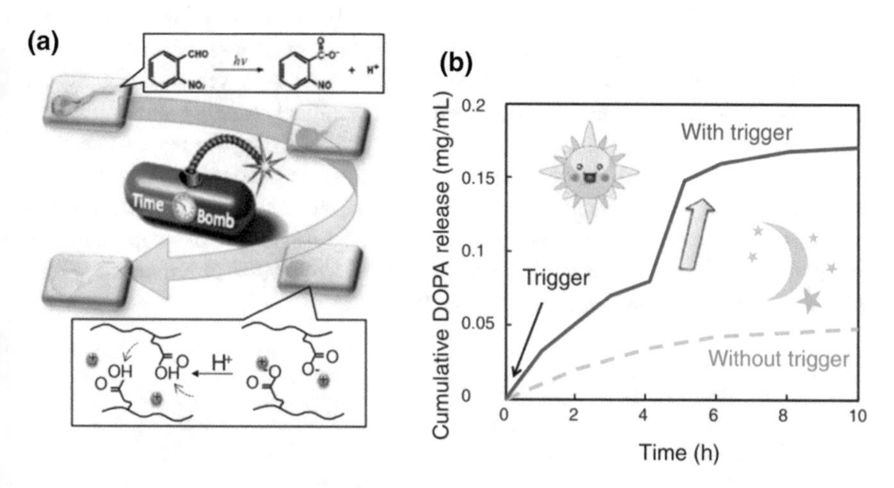

Fig. 6 **a** Release of the timed explosive drug by the spatial pH-jump reaction from pH-responsive hydrogel triggered by light **b** effect of light on time-dependent drug release of hydrogel [39]

7 Enzyme-Responsive SAP

These are the new class of stimuli-responsive SAPs. The precise selectivity and high efficiency of enzymatic catalysis in enzyme-responsive polymers with intriguing properties distinguish this SAP from other smart polymers. This class of polymeric materials changes its structure and functionality with response to direct action of the enzyme even undermild conditions. Here the response is towards enzymes including lipases, proteases, phosphatases, kinases, glycosidases, acyl transferases, and redox enzymes. Enzyme-responsive polymers change their physical and chemical properties in a reversible or irreversible way in response to the enzymes. Externally applied stimuli including pH, temperature, light, solvent polarity, electric/magnetic field, and small (bio-)molecules cause changes in a materialmay not be compatible with external application in a biological environment because the changes are not very targeted or specific to the material and may also affect other components of the biological environment. A variety of naturally occurring enzymes present in the body offer additional benefits over external stimuli such as pH and temperature as they are supplied by the biological environment itself. Here the addition of external stimuli is not needed as long as the naturally occurring enzyme matches the trigger provided by the enzyme of the responsive material [40]. Reactions that are catalysed by enzymes are highly effective and selective toward particular substrates. The intricate properties are triggered by the integration of enzyme-catalyzed reactions with responsive polymers. Responsive polymers can exist in a variety of states, such as solutions, gels, self-assembled aggregates, (multilayer) films, and bulk solids, and they can undergo reversible or irreversible changes to their chemical structure and/or physical characteristics [41]. Due to the inherent bio-compatibility, enzyme-responsive polymers have application in various field including drug or gene nanocarriers, regenerative medicine, diagnostics, smart actuators, adaptive coatings, and self-healing materials. The enzyme-responsive polymer can be formed either by the ezyeme-degradable polymer or by modifying it by grafting with other polymers to get the specific enzyme-responsive property (Fig. 7).

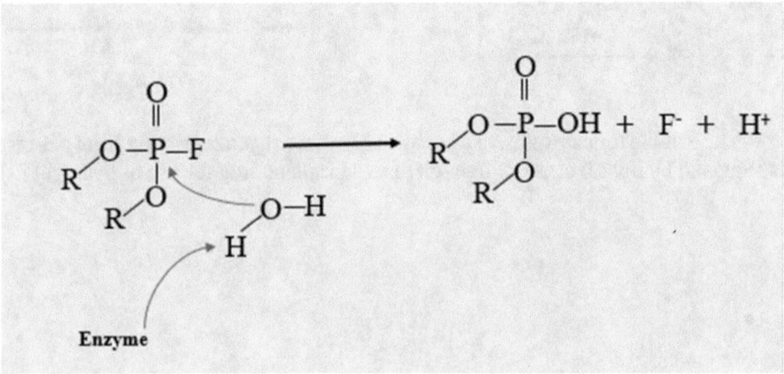

Fig. 7 Enzymatic action on enzyme-responsive SAP

8 CO_2-Responsive SAP

These are the recently developed SAPs which switch their properties from the relatively hydrophobic state to the hydrophilic state in response to CO_2 availability. Here CO_2 can be readily used as a stimulus due to its easy availability, relatively benign and non-accumulating property. Again this can be removed and recycled easily. In the mechanism involved here is the presence of CO_2, SAP gets protonated by holding some amount of water and when CO_2 is removed it is collapsed back by getting deprotonated. A study shows the swelling of the microgel inside water in the presence of CO_2 and when the concentration increases above certain threshold concentration, microgels were irreversibly in a collapsed structure. Another study shows the swelling action in the presence of CO_2 for the first time where the hydrogel swells in the presence of CO_2 and shrinks rapidly when immersed in tap water and slowly when exposed to air. The swelling action of CO_2 responsive SAP is more in carbonated water than in deionized or tap water [42]. The CO_2 responsive SAP was synthesised from grafting either of N-[3-(dimethylamino)propyl]-methacrylamide (DMAPMAm) or 2-N-morpholinoethyl methacrylate (MEMA) with N,N′-dimethylacrylamide (DMAAm), (as monomer and a self-cross-linker) in the presence of CO_2 [43] (Fig. 8).

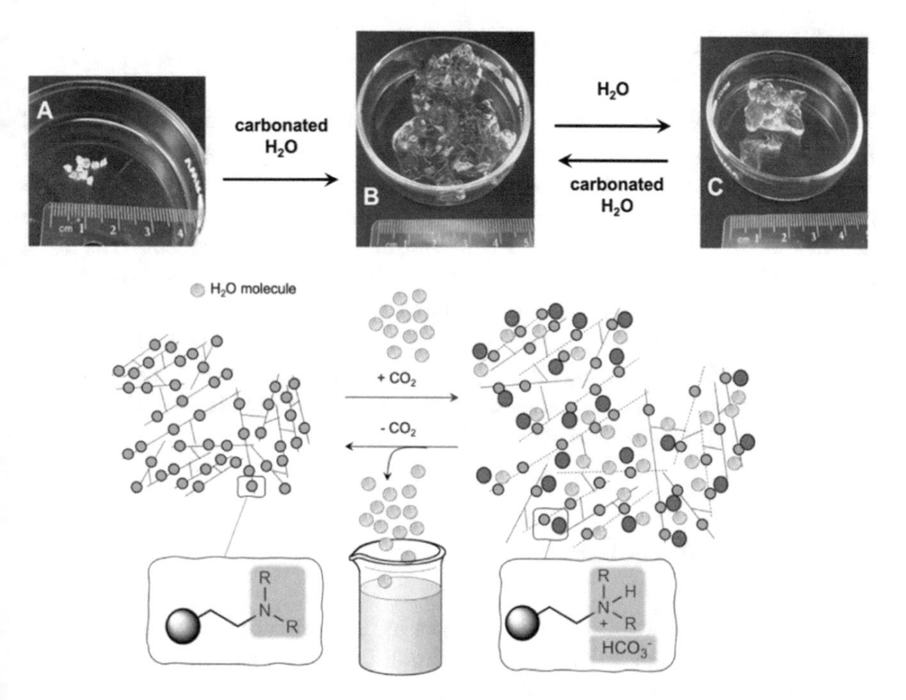

Fig. 8 Representation of swelling/deswelling action of CO_2-responsive SAP in the presence and absence of CO_2, respectively. The dashed lines represent swollen SAP (right)

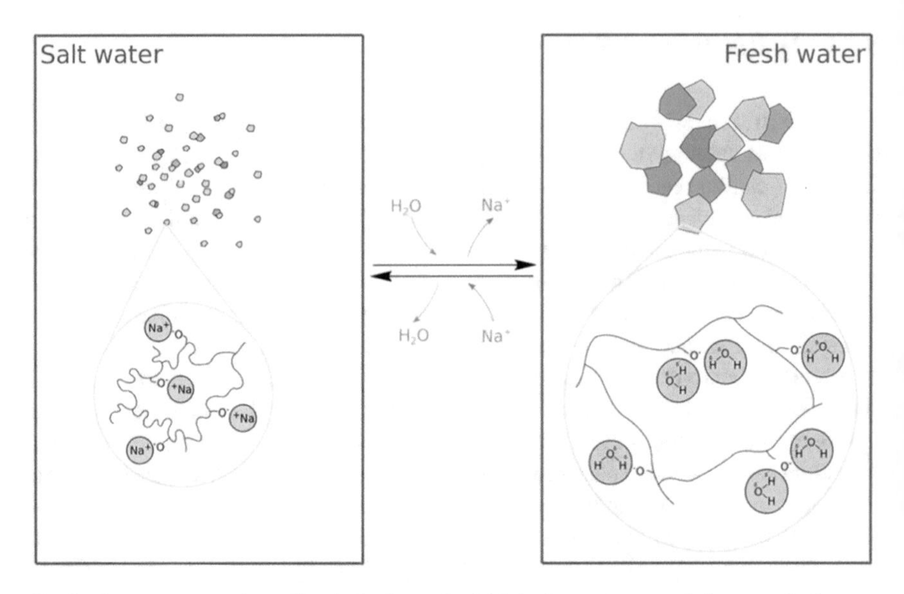

Fig. 9 The structure and swelling behaviour of a SAP in the presence and absence of salt water respectively

9 Salt-Water Responsive SAPs

These are the new generation of SAPs where the polymer structure changes in response to salt water. In the presence of salt water the SAP gets collapsed and when removed from that environment and placing it in normal water it again starts to swell. This is mostly used in robotics as actuators or hydrogel-based artificial muscles where swelling of the gels leads to the contraction of the muscle. The novel actuator is formed by the encapsulated SAP gel within discrete cells and the absorption of water takes place through the membrane wall. The phenomena of contraction takes place through the swelling action, by which a significant amount of pressure is generated resulting in a change in shape. A study developed a new salt-responsive marine soft robot from a superabsorbent polymer-based Bubble Artificial Muscle (SAP-BAM) [44] (Figs. 9 and 10).

10 Bi/Multi Responsive SAPs

When the Super absorbent polymer changes its chemical or physical properties in response to the change of two or more stimuli then the Smart SAP comes under the bi or multi-responsive category. These SAPs include any two or more stimuli such as Temp, pH or electric–magnetic field. A temperature and pH responsive SAP synthesized from modified Maleic anhydride possess a strong swelling capacity at

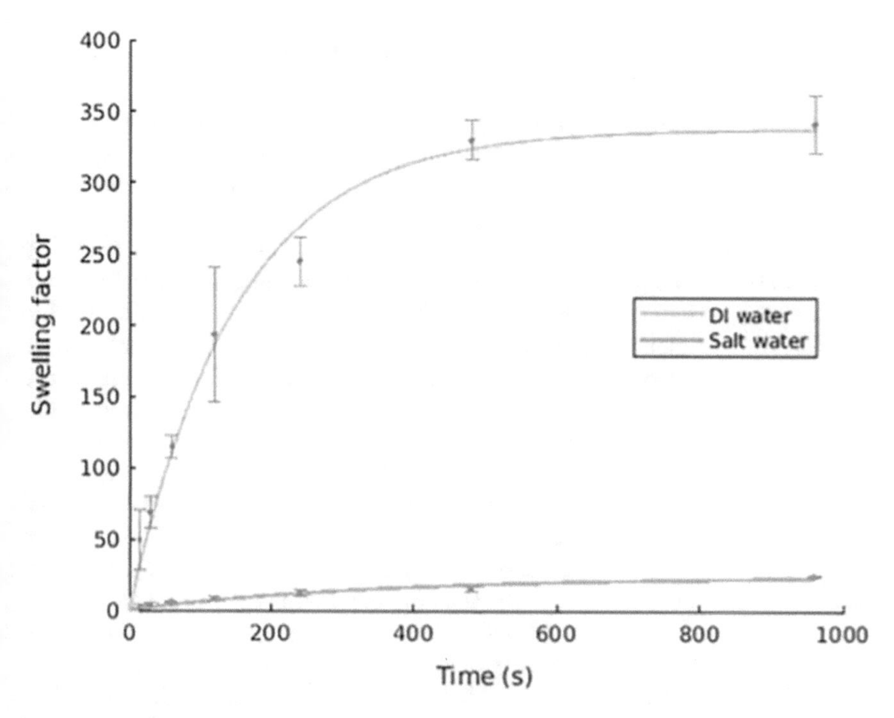

Fig. 10 The swelling behaviour of sodium polyacrylate powder in deionised and salt water. Error bars shows one standard deviation [44]

a higher temperature including two types of diffusion [45]. Another Temperature-Sensitive and Magnetic Properties Employed a Superabsorbent Polymer with High Swelling Ratio found as an Efficient Dewatering Medium of Fine Coal [46] (Fig. 11).

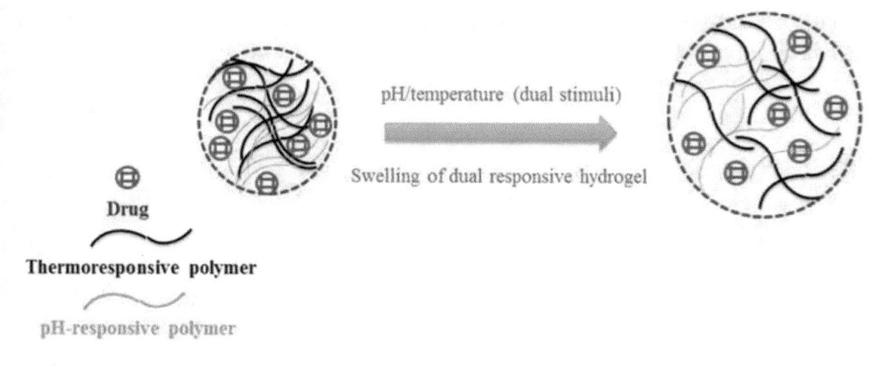

Fig. 11 Dual-responsive SAP used in drug release [47]

11 Other Bio-based Superabsorbent Polymers

To overcome the issues faced in the case of petroleum-based SAPs, bio-based SAPs from polysaccharides (starch, cellulose, chitosan carrageenan, alginate, and gum arabic), proteins (soybean, fish and collagen-based) and gelatin have been investigated as an alternative. These SAPs are readily available, biocompatible, biodegradable, non-toxic, renewable and sustainable and cost-effective. Certain properties such as water solubility of alginate and thermo-sensitivity of chitosan make them a suitable polymer for various applications. Grafting of protein with acrylic acid has biomedical application by inducing cell-interactive properties. Polysaccharides have gained focus due to the limitation of using proteins. Polysaccharides can be produced by bacteria including bacterial hyaluronan, gellan or xanthan and also from plants and animals by bio-synthesis. Due to presence of functional groups such as alcohols, carboxylic acids in water-soluble polysaccharides, these can be cross-linked or grafted with other polymers for advanced applications. Sodium alginate extracted from brown algae is an anionic water-soluble polysaccharide. The carboxylic acid groups present in alginate gets negatively charged in aqueous solution leading to the pH-responsive behaviour in SAPs and have applications in biomedical field for controlled drug release, cell encapsulation, dental impression, wound dressing) and in bioplastics as packaging, textiles, paper and in food industry as a stabilizer, emulsifier and gelling agent. The shortcoming of bio-based SAPs such as their low solubility, and the complicated purification process leads to the modification of these SAPs to develop semi-synthetic SAPs. Carrageenan is a bio-based SAP refined from the cell wall of red algae that has potential use in pharmaceutical and food industry as emulsifier, stabilizer or thickeners. Another researcher uses Agarose, a bio-based polymer extracted from red algae with covalently crosslinked alginate or chitosan using carbonyldiimidazole to study the effect of charges on neural tissue scaffolds. A bio-based SAP with excellent bio-compatibility and water-soluble nature was prepared by the cross-linking of poly-itanoic acid using potassium persulfate as an initiator and poly(ethylene glycol) diacrylate as a cross-linking agent by free-radical polymerization [32, 48].

References

1. Laftah, W.A., Hashim, S., Ibrahim, A.N.: Polymer hydrogels: a review. Polym.-Plast. Technol. Eng. **50**(14), 1475–1486 (2011)
2. Wu, L., Liu, M., Liang, R.: Preparation and properties of a double-coated slow-release NPK compound fertilizer with superabsorbent and water-retention. Biores. Technol. **99**(3), 547–554 (2008)
3. Bakass, M., Mokhlisse, A., Lallemant, M.: Absorption and desorption of liquid water by a superabsorbent polymer: effect of polymer in the drying of the soil and the quality of certain plants. J. Appl. Polym. Sci. **83**(2), 234–243 (2002)
4. Sannino, A., Esposito, A., Rosa, A.D., Cozzolino, A., Ambrosio, L., Nicolais, L.: Biomedical application of a superabsorbent hydrogel for body water elimination in the treatment of edemas.

J. Biomed. Mater. Res. Part A Official J. Soc. Biomater. Jpn. Soc. Biomater. Aust. Soc. Biomater. Korean Soc. Biomater. **67**(3), 1016–1024 (2003)
5. Kabiri, K., Omidian, H., Zohuriaan-Mehr, M.J., Doroudiani, S.: Superabsorbent hydrogel composites and nanocomposites: a review. Polym. Compos. **32**(2), 277–289 (2011)
6. Chen, F., Tillberg, P.W., Boyden, E.S.: Expansion microscopy. Science **347**(6221), 543–548 (2015)
7. Gao, R., Yu, C.C.J., Gao, L., Piatkevich, K.D., Neve, R.L., Munro, J.B., ... Boyden, E.S.: A highly homogeneous polymer composed of tetrahedron-like monomers for high-isotropy expansion microscopy. Nat. Nanotechnol. **16**(6), 698–707 (2021)
8. Yang, L., Liu, H., Ding, S., Wu, J., Zhang, Y., Wang, Z., ... Tao, G.: Superabsorbent fibers for comfortable disposable medical protective clothing. Adv. Fiber Mater. **2**(3), 140–149 (2020)
9. Zhang, Z., Huang, G.: Micro-and nano-carrier mediated intra-articular drug delivery systems for the treatment of osteoarthritis. J. Nanotechnol. **2012** (2012)
10. Hosseini, M., Makhlouf, A.S.H. (eds.).: Industrial Applications for Intelligent Polymers and Coatings, vol. 1, pp. 1–710. Springer, Berlin/Heidelberg, Germany (2016)
11. Aguilar, M.R., Elvira, C., Gallardo, A., Vázquez, B., San Román, J.: Smart polymers and their applications as biomaterials. In: Ashammakhi, N., Reis, R.L., Chiellini, E. (eds.) Topics in Tissue Engineering, Chapter 6, 27 pp., e-book. Expertissues (2007)
12. Stryer, L.: Biochemistry. Freeman, New York (1999)
13. Roy, D., Cambre, J.N., Sumerlin, B.S.: Future perspectives and recent advances in stimuli-responsive materials. Prog. Polym. Sci. **35**, 278–301 (2010). https://doi.org/10.1016/j.progpolymsci.2009.10.008
14. Hoogenboom, R.: Temperature-responsive polymers: properties, synthesis, and applications. In: Smart Polymers and Their Applications, pp. 13–44. Woodhead Publishing (2019)
15. Oladosu, Y., Rafii, M. Y., Arolu, F., Chukwu, S. C., Salisu, M. A., Fagbohun, I. K., ... & Haliru, B. S.: Superabsorbent polymer hydrogels for sustainable agriculture: A review. Hortic **8**(7), 605 (2022)
16. Ishida, K., Uno, T., Itoh, T., Kubo, M.: Synthesis and property of temperature-responsive hydrogel with movable cross-linking points. Macromolecules **45**(15), 6136–6142 (2012)
17. Qiu, Y., Park, K.: Environment-sensitive hydrogels for drug delivery. Adv. Drug Deliv. Rev. **53**(3), 321–339 (2001)
18. Carreira, A.S., Gonçalves, F.A.M.M., Mendonça, P.V., Gil, M.H., Coelho, J.F.J.: Temperature and pH responsive polymers based on chitosan: applications and new graft copolymerization strategies based on living radical polymerization. Carbohyd. Polym. **80**(3), 618–630 (2010)
19. Katono, H., Maruyama, A., Sanui, K., Ogata, N., Okano, T., Sakurai, Y.: Thermo-responsive swelling and drug release switching of interpenetrating polymer networks composed of poly (acrylamide-co-butyl methacrylate) and poly (acrylic acid). J. Control. Release **16**(1–2), 215–227 (1991)
20. James, H.P., John, R., Alex, A., Anoop, K.: Smart polymers for the controlled delivery of drugs–a concise overview. Acta Pharmaceutica Sinica B **4**(2), 120–127 (2014)
21. Pourjavadi, A., Barzegar, S.: Synthesis and evaluation of pH and thermosensitive pectin-based superabsorbent hydrogel for oral drug delivery systems. Starch-Stärke **61**(3–4), 161–172 (2009)
22. Miguel, S.P., Ribeiro, M.P., Brancal, H., Coutinho, P., Correia, I.J.: Thermoresponsive chitosan–agarose hydrogel for skin regeneration. Carbohyd. Polym. **111**, 366–373 (2014)
23. Hoogenboom, R., Thijs, H.M., Jochems, M.J., van Lankvelt, B.M., Fijten, M.W., Schubert, U.S.: Tuning the LCST of poly (2-oxazoline) s by varying composition and molecular weight: alternatives to poly (N-isopropylacrylamide)? Chem. Commun. **44**, 5758–5760 (2008)
24. Medeiros, S.F., Santos, A.M., Fessi, H., Elaissari, A.: Stimuli-responsive magnetic particles for biomedical applications. Int. J. Pharm. **403**(1–2), 139–161 (2011)
25. Aguilar, M.R., San Román, J.: Introduction to smart polymers and their applications. In: Smart Polymers and Their Applications, pp. 1–11. Woodhead Publishing (2019)
26. Liu, J., Huang, Y., Kumar, A., Tan, A., Jin, S., Mozhi, A., Liang, X.J.: pH-sensitive nano-systems for drug delivery in cancer therapy. Biotechnol. Adv. **32**(4), 693–710 (2014)

27. Irfan, J., Hussain, M.A., Haseeb, M.T., Ali, A., Farid-ul-Haq, M., Tabassum, T., ... Naeem-ul-Hassan, M.: A pH-sensitive, stimuli-responsive, superabsorbent, smart hydrogel from psyllium (Plantago ovata) for intelligent drug delivery. RSC Adv. **11**(32), 19755–19767 (2021)
28. Vardanyan, R., Hruby, V.: Synthesis of Essential Drugs. Elsevier (pH-responsive) (2006)
29. Gruyaert, E., Debbaut, B., Snoeck, D., Díaz, P., Arizo, A., Tziviloglou, E., ... De Belie, N.: Self-healing mortar with pH-sensitive superabsorbent polymers: testing of the sealing efficiency by water flow tests. Smart Mater. Struct. **25**(8), 084007 (2016)
30. Osada, Y., Hasebe, M.: Electrically activated mechanochemical devices using polyelectrolyte gels. Chem. Lett. **14**(9), 1285–1288 (1985)
31. Zohuriaan-Mehr, M.J.: Super-absorbents. Iran Polym. Soc. Tehran **228**, 2–4 (2006)
32. Mignon, A., De Belie, N., Dubruel, P., Van Vlierberghe, S.: Superabsorbent polymers: a review on the characteristics and applications of synthetic, polysaccharide-based, semi-synthetic and 'smart' derivatives. Eur. Polymer J. **117**, 165–178 (2019)
33. Bünsow, J., Johannsmann, D.: Electrochemically produced responsive hydrogel films: influence of added salt on thickness and morphology. J. Colloid Interface Sci. **326**(1), 61–65 (2008)
34. Kulkarni, R.V., Sa, B.: Electroresponsive polyacrylamide-grafted-xanthan hydrogels for drug delivery. J. Bioact. Compat. Polym. **24**, 368–384 (2009)
35. Murdan, S.: Electro-responsive drug delivery from hydrogels. J. Control. Release **92**, 1–17 (2003)
36. James, H.P., John, R., Alex, A., Anoop, K.R.: Smart polymers for the controlled delivery of 1232 drugs—a concise overview. Acta Pharmaceutica Sinica B **4**(2), 120–127 (2014)
37. Savina, I.N., Galaev, I.Y., Mikhalovsky, S.V.: Smart polymers for bioseparation and other biotechnological applications. In: Smart Polymers and their Applications, pp. 533–565. Woodhead Publishing (2019)
38. Mudiyanselage, T.K., Neckers, D.C.: Photochromic superabsorbent polymers 1. Soft Matter **4**(4), 768–774 (2008)
39. Techawanitchai, P., Idota, N., Uto, K., Ebara, M., & Aoyagi, T.: A smart hydrogel-based time bomb triggers drug release mediated by pH-jump reaction. Sci Tech Adv Mater (2012)
40. Asha, A.B., Srinivas, S., Hao, X., Narain, R.: Enzyme-responsive polymers: classifications, properties, synthesis strategies, and applications. In: Smart Polymers and their Applications, pp. 155–189. Woodhead Publishing (2019)
41. Hu, J., Liu, S.: Engineering responsive polymer building blocks with host–guest molecular recognition for functional applications. Acc. Chem. Res. **47**(7), 2084–2095 (2014)
42. Jansen-van Vuuren, R.D., Drechsler Vilela, G., Ramezani, M., Gilbert, P.H., Watson, D., Mullins, N., ... Jessop, P.G.: CO_2-responsive superabsorbent hydrogels capable of >90% dewatering when immersed in water. ACS Appl. Polym. Mater. **3**(4), 2153–2165 (2021)
43. Riabtseva, A., Ellis, S.N., Champagne, P., Jessop, P.G., Cunningham, M.F.: CO_2-responsive branched polymers for forward osmosis applications: the effect of branching on draw solute properties. Ind. Eng. Chem. Res. **60**(27), 9807–9816 (2021)
44. Gosden, D., Diteesawat, R.S., Studley, M., Rossiter, J.: Saltwater-responsive bubble artificial muscles using superabsorbent polymers. Front. Robot. AI **9**, 960372 (2022)
45. Mazi, H., Surmelihindi, B.: Temperature and Ph-sensıtıve super absorbent polymers based on modified maleic anhydride. J. Chem. Sci. **133**(1), 1–14 (2021)
46. Zhang, S., Chen, H., Liu, S., Guo, J.: Superabsorbent polymer with high swelling ratio, and temperature-sensitive and magnetic properties employed as an efficient dewatering medium of fine coal. Energy Fuels **31**(2), 1825–1831 (2017)
47. Chatterjee, S., & Hui, P. C. L.: Stimuli-responsive hydrogels: An interdisciplinary overview. Hydrogels—smart materials for biomedical applications, 1–23 (2018)
48. Choi, H., Park, J., Lee, J.: Sustainable bio-based superabsorbent polymer: poly(itaconic acid) with superior swelling properties. ACS Appl. Polym. Mater. (2022)

Recycling and Reuse of Superabsorbent Polymers

Ankita Subhrasmita Gadtya, Debajani Tripathy, and Srikanta Moharana

Abstract Super-absorbent polymers (SAPs) are three-dimensional network polymers that can absorb hundreds of times their own weight in distilled water and are either organic or inorganic. They are extensively employed in many different industries, including those that deal with agriculture, biomedicine, everyday physiological goods, separation technology, and wastewater treatment. Polysaccharides are increasingly being combined with synthetic monomers to create semi-synthetic, hybrid SAPs for specialized applications requiring finely tuned properties, such as wound dressings, fertilizers, or self-healing concrete, where the low solubility of synthetic polymers, purification issues, or the need for organic solvents are significant barriers to entry. Products made of polypropylene, polyethylene, elastics, cellulose, and superabsorbent polymers (SAPs) are among the billions that end up in landfills every year. The chapter discusses the development of superabsorbent polymers since 1961, along with the polymerisation procedures for these materials, their recycling, and their uses.

Keywords Superabsorbent polymer · Recycle · Reuse · Synthesis · Classification

1 Introduction

Superabsorbent polymers are utilised to maintain their original form (granular, fibre, sheets, etc.) after becoming saturated with water. A common combination of acryl amide, acrylic acid, and their corresponding salts is used to create superabsorbent polymers (SAPs). SAPs can be either natural or manufactured (based on chemicals) (based on polysaccharides and polypeptides) [1, 2]. Due to the variety in kind and quality of the different SAP products on the market, costs per kilogramme range from US$10 to $40. These substances can decrease bulk density and raise plant available water (PAW) (BD) [3]. This study aims to evaluate how the sustainability

A. S. Gadtya · D. Tripathy · S. Moharana (✉)
School of Applied Sciences, Centurion University of Technology and Management, R.Sitapur, Odisha, India
e-mail: srikantanit@gmail.com; srikanta.moharana@cutm.ac.in

S. Pradhan and S. Mohanty (eds.), *Bio-based Superabsorbents*, Engineering Materials,
https://doi.org/10.1007/978-981-99-3094-4_9

of disposable newborn diapers may be enhanced by instituting a circular economy for the precious materials—specifically the superabsorbent polymers (SAPs)—that contribute significantly to their initial environmental consequences [4]. Using a life cycle evaluation to evaluate the impact of disposable diapers is nothing new (LCA). A classic example of life cycle assessment is the evaluation of disposable vs reusable diapers. Consumers were allowed to consider costs and benefits when deciding between disposable and cloth diapers. The primary environmental effect of cloth diapers is the energy and water consumed in the laundering process [5]. Manufacturers have made great efforts in reducing the environmental effect of disposable diapers via improved product design [6]. For instance, in the last two decades, manufacturers have managed to reduce the weight of disposable diapers by about half. The usage of SAPs in contact lenses began in the 1950s, and since then, the material has found a wide variety of applications, leading to 2005 production levels of about 1.5 million metric tones [2]. SAP content has grown from 0.7 g in 1987 to 11.1 g in 2011 while the weight of practically all other diaper materials has decreased [7]. Life cycle implications of disposable diapers, including as greenhouse gas emissions, water usage, and other air pollutants, must be addressed if the increasing environmental impacts of the product are to be mitigated [8]. Further, breakthroughs in product materials may not be as rapid to come about as technological advancements in recycling and reducing the waste of high-value materials at the end of their useful lives [9].

There has been some progress in developing alternatives to the widespread practice of throwing away used diapers in an effort to reduce this waste of resources. The plastic components of a diaper seem to be the main focus of recycling programmes, or their incineration for energy recovery [10]. High manufacturing burdens should be the focus of recycling activities in order to overcome these challenges as well as provide environmental benefits [11]. Due to their high economic and feedstock value and their significant role in contributing to greenhouse gas emissions from diapers, SAPs provide a potential for reuse [7, 12]. Repurposing SAP in its original context has the ability to save production costs and opens up a new direct market. The current SAP recycling and recovery efforts are insufficient [13]. The topic of when SAP can be successfully recovered and reused becomes increasingly relevant as the technology to do so mature. In light of the ambiguities surrounding SAP recovery and reuse, this article examines the environmental advantages and disadvantages of a possible SAP recovery process in an effort to pinpoint the threshold at which SAP recovery may lead to reduced manufacturing loads and material recuperation.

2 Overview of Super Absorbing Materials

Superabsorbent polymers (SAPs), also known as hydrogels, are networks of flexible polymer chains with dissociated ionic functional groups that are loosely cross-linked in three dimensions to avoid dissolution [14]. These substances are like water, electrolyte solutions, synthetic urine, brines, biological fluids like urine, sweat, and blood that can absorb liquids up to 15 times their own dried weight, under load or no load. SAPs can hold a lot of water in their pores while keeping their shape even after swelling due to their ionic structure via capillary action and under pressure. When the size of the interior pores shrinks, water is discharged. The promotion of diffusion of water in the polymeric network is done by the ionic functional group in the polymeric chain [15, 16].

The elasticity of the hydrogel, which is a result of its network structure, is what reduces absorption power. Furthermore, because they only immobilise the fluid via trapping rather than by keeping it in the structure, the absorbed fluid is difficult to release. They are cross-linked polyelectrolytes that are hydrophilic polymers with carboxylic acid, carboxyl amide, hydroxyl amine, and imide groups that are insoluble in water. The polymeric chain typically has ionic functional groups to aid in the diffusion of water throughout the network. After water absorption and swelling, the SAP particle shape (granule, fibre, film, etc.) must essentially be maintained, meaning that the swelled gel strength needs to be high enough to prevent a loosening, mushy, or slimy state. SAPs have the potential to alter the dynamics of evaporation because of their capacity to increase soil water retention [17, 18]. SAPs are widely employed in numerous industries, including sanitary products, medicine, horticulture, and agriculture, as a result of their exceptional qualities. In addition to this, SAPs are also used in numerous other applications like paper towels, surgical sponges, meat trays, disposable mats for outside gateways and in restrooms, household pet litter, bandages, and wound dressings, the fundamental property of water absorption has suggested the usage. The ability of the swollen gels to release the water to the environment as vapour has also been used in a variety of ways, such as soil conditioners or items that regulate humidity. SAPs can also be utilised to release water-soluble materials into the environment from inside the network structure.

3 Classification of Superabsorbent Polymer Materials (SAPs)

SAPs are of many types based on different aspects. On the basis of presence or absence of electrical charge on the cross-linked chains, it is mainly of the following four types [19, 20];

(i) Non-ionic (neutral): This type of SAP absorbs the water and other aqueous fluids by combining aqueous fluids with the hydrophilic groups found along the polymer chain through energetic and entropic interactions. The water molecules here are solvated through hydrogen bonding. Polyvinyl alcohol (PVA) gels are best examples of non-ionic SAP which are prepared by monomeric reaction of vinyl acetate. The hydroxylmethyl cellulose-based SAP are also examples of this kind of SAP [21].

(ii) Ionic (cationic/anionic): Ionic SAPs are also known as ionic hydrogels or ionic hydrocolloids. The ionic hydrocolloids are typically cross-linked ionic polymers that can absorb and hold onto pure water under mild external pressure in amounts at least ten times greater than their dry weight. As different groups connected to the polymer chain can easily become ionic, these SAPs are electrically charged in solution. Ionic SAPs can be cationic or anionic in nature. They have positive or negative charges in the backbone of their structure. The examples of cationic SAPs are partially neutralized polyamine and that of the anionic SAPs are acylated based or sulfonate based SAP [22].

(iii) Amphoteric: Three-dimensional cross-linked networks of anionic, cationic, and neutral repeating units make up amphoteric SAPs. The network's ionic units produce an osmotic pressure differential between the swelling media and the SAP. The SAP swells as a result of a significant water flow into the network to balance the osmotic pressure imbalance. Depending on the makeup of the individual repeat units, amphoteric SAPs may have a net positive, negative, or neutral charge. The type and quantity of the amphoteric SAP's net charge determines its swelling properties and its capacity to absorb charged species from aqueous solutions, such as dyes and metal ions. One of the example of amphoteric SAP is poly (SA-co-METAC) which is prepared by the anionic monomer sodium acrylate (SA) and the cationic monomer of [2-(methacryloyloxy)ethyl]trimethylammonium chloride (METAC). This SAP is prepared (Fig. 1) in the presence of one of the cross-linking agent N,N′-methylenebisacrylamide (MBA) [23].

Zwitter ionic: The zwitter-ionic based SAP has a different cationic and anionic group in its structural unit. In an experiment by Rehman et al. the authors have reported a zwitter-ionic based SAP hydrogel (ZI-SAH) prepared via free radical synthesis technology. This ZI-SAH was helpful in removal of organic dyes like crystal violet (CV) and Congo red (CR) from the aqueous medium. The removal efficiency was found to be more than 75% after multiple cycles of being used. The behavior of ZI-SAH was also semi-solid, pseudo-plastic and non-Newtonian in nature and all these parameters indicated towards application of it in the adsorption [24]. Zwitter-ionic Poly(Sulfobetaine Methacrylate) (PSBMA) was prepared with the help of electrospinning methodology. The electrospun PSBMA showed excellent water absorption capacity due to which it acts as a super absorbent polymer. The water absorption capacity was found to be 353% (w/w) of the electospun PSBMA [25]. Another example included zwitterionic poly(sulfobetaine methacrylate) whose synthesis method in Fig. 2 was given below [21].

Apart from these four types; SAPs are also classified into another three types based on their origin of polymer [26]. They include;

(i) Synthetic polymer based SAP: Synthetic polymer based SAPs are employed in a wide range of biomedical and non-biomedical applications, including electrophoresis gels, diapers manufacturing, water purification systems, drug delivery systems, and more recently, self-healing applications. However,

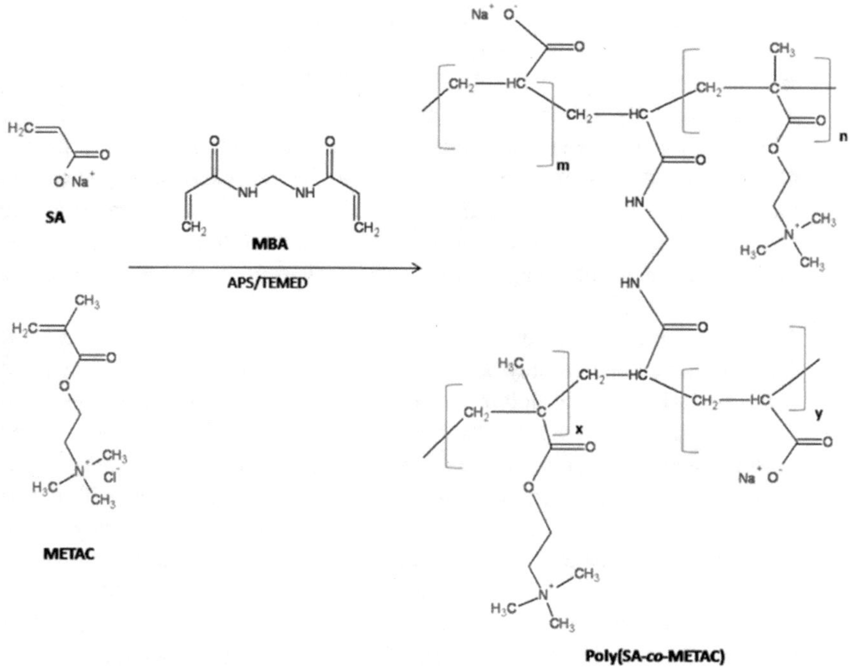

Fig. 1 Synthesis of amphoteric SAP Poly (SA-co-METAC) [23]

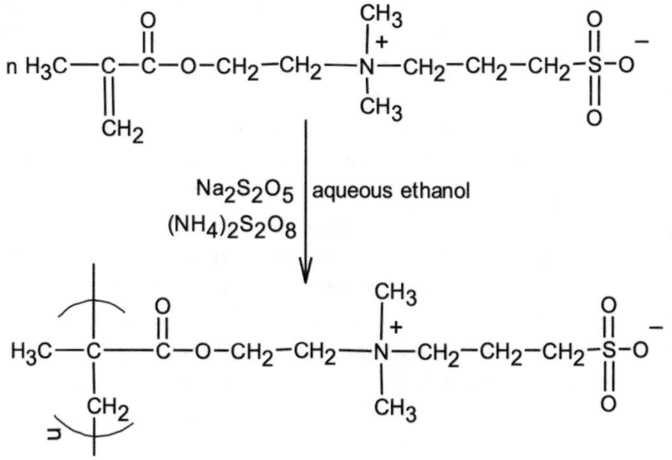

Fig. 2 Synthesis of zwitterionic poly(sulfobetaine methacrylate) [21]

despite the fact that synthetic SAPs are widely used, they are neither sustainable nor renewable and frequently lack components that degrade into the environment. Examples of synthetic polymer based SAP are cross-linked sodium polyacrylate, sodium poly(vinyl sulfonate), and poly(ethylene glycol) etc. Although all hydrophilic monomers, including nonionic monomers like ethylene glycol, can be thought of as monomers for SAP, sodium acrylate or acrylic acid is the monomer that is currently most typically employed. These are used due to their easy synthetic method for large molecular weight polymers and as the polymers have a high charge density, acrylic acid or sodium acrylate that is highly favored [27–29].

(ii) Natural polymer based SAP: There are two types of SAPs in this category: (a) cross-linked natural polymers having charges on the polymeric chain, and (b) cross-linked natural polymers modified via the substitution processes to add low molecular weight substituents with charges. Cross-linking can be accomplished by coupling radicals produced by electron or gamma ray irradiation or by reacting functional groups on various polymer chains with multifunctional reagents like glutaraldehyde and ethylene glycol diglycidyl ether. Different polysaccharides like chitosan, hyaluronic acid etc.; poly(amino acid) such as poly(glutamic acid), poly(aspartic acid) etc.; proteins like collagen etc. are being used as SAP which are naturally derived. The natural SAPs have low cytotoxicity, good biodegradability and high drug binding capacity. A dual stimuli responsive natural polymer based superabsorbent hydrogel was prepared through the cross-linking process. A highly porous grafted polymer soy protein isolate (SPI)/grafted[2-(4-((acrylamido)methyl)-1H-1,2,3-triazol-1-yl)-4-vinylbenzoic acid-co-4-(4-hydroxyphenyl)butanoic acid] [SPI-g-(ATVBA-co-HPBA)-g-SPI] was prepared [30].

(iii) Natural/synthetic hybrid polymer based SAP: This type of SAP primarily consists of polysaccharide-graft-polyacrylates. Polysaccharides are exciting resources for SAP raw materials since they are the most widely distributed, reasonably priced, and biodegradable naturally occurring polymer. There have been reports of several polysaccharide-graft-polyacrylics made from starch, cellulose, chitin, chitosan, guar gum, xanthan, and alginate. Starch–graft–poly(acrylic acid) (St-g-PAA) was the first natural synthetic hybrid SAP. Gok et al. have synthesized wheat starch grafted PAA (WS-g-PAA) by using L-cysteine hydrochloride monohydrate (CyS) and thioglycolic acid (TGA). The mucoadhesive property of the WS-g-PAA is due to the thiolation [31]. In recent times super water absorbents (SWA) based on polysaccharides/poly(acrylic acid) (PAAc) were prepared with the help of gamma ray radiation technique by Relleve et al. The different SWAs' gel fractions varied from 31 to 97%, and the degree of swelling peaked at around 5890 g H_2O/g dry gel. As the starch was naturally derived from cassava, the biodegradability was 42% within 85 days only. Also it was observed that the water retention capacity was more than 20 days in the sandy soil. The absorption of water continued after 62 days also [32]. In another experiment St-g-PAA hydrogels were produced by designing the experimental method. Because Poly(acrylic acid) (PAA) is more hydrophilic, more polar, and more soluble than other vinyl monomers, it can be categorised as a novel hydrogel in St-based hydrogels [33].

On the basis of the composition of polymer again SAPS are of three types. They are [20, 34];

(i) Homopolymeric: The polymeric networks of these kinds of SAPs are made from a single species of monomeric unit. Depending on the type of monomer and the polymerisation process, homopolymers may have a cross-linked skeletal structure.
(ii) Copolymeric: These are made up of two or more distinct monomer species that each contain at least one hydrophilic component and are placed randomly, in blocks, or alternately throughout the polymer network's chain. Preparation of SAP was done by the cross-linking of acrylic acid (AA) and acrylamide (AM) copolymers. The cross-linked co-polymer of AA and AM (CCPAA) can absorb up to 825 gm of water within 10 min [35].
(iii) Multipolymer interpenetrating polymeric hydrogel: They are constructed of two synthetic or natural polymers that have been cross-linked individually and are combined into a network. The chemical cross-linking and polymerisation process was done through (xanthum gum-poly vinyl alcohol) interpenetrating polymer hydrogel [(XG-PVA)IPN] to get the biodegradable and biocompatible PAA/(XG-PVA)IPN in the presence of the monomeric solution. The equilibrium water absorbency capacity and water retention ability of the superabsorbent polymers based on biodegradable natural polymers were improved. This synthesized SAP has been used in novel drug delivery system [36].

4 Chemical Synthesis Methods

There are different chemical synthesis methods for the preparation of SAPs. Few of them are discussed below [14, 37].

4.1 Bulk Polymerisation

Bulk polymerisations in other words are known as mass polymerisation. It is one of the simplest method in which monomers are polymerised without the need of solvents or dispersants using initiators, light, heat, or radiation. The high monomer concentration leads to a high rate and degree of polymerisation. However, the reaction viscosity increases substantially, and the heat produced by polymerisation is difficult to dissipate. In the event of producing a lot of heat, these issues can be avoided by employing the lower temperature and low concentration of initiators. Bulk polymerisation has the benefit of producing high-purity, high-molecular-weight polymers. This method is used to create polyacrylate SAPs.

4.2 Solution Polymerisation

Ionic or neutral monomers are combined with the multifunctional cross-linking agent in solution co-polymerisation/cross-linking reactions. Thermal, UV, or redox initiator systems can all be used to start the polymerisation process. The main benefit of solution polymerisation over bulk polymerisation is the presence of solvent acting as a heat sink. For complete purification the prepared SAPs should be washed several times with distilled water for the removal impurities like unreacted monomers, cross-linked agents, initiators, oligomers etc. When the amount of water during polymerisation is more than the water content corresponding to the equilibrium swelling, phase separation takes place and the heterogeneous SAP is created. One of the best example is the preparation of poly(2-hydroxy ethyl methacrylate) SAPs utilising ethylene glycol dimethacrylate as a cross-linking agent and hydroxyl ethyl methacrylate as the starting material. These SAPs are the water-swelling resin which control the density of cross-linking to absorb the desired amount of water. The poly(acrylamide-co-acrylic acid), a salt resistant super absorbent, was prepared by Z. Chen et al. with the help of the solution polymerisation method [38]. The copolymer was surface-cross linked using ethylene glycol diglycidyl ether to improve the strength, resilience, and dispersion of the swelling hydrogel. The surface-cross linked copolymer was then blended with aluminium sulphate ($Al_2(SO_4)_3$) and sodium carbonate (Na_2CO_3) to produce the desired sample to get enhanced properties. In another experiment the SAP hydrogel samples were prepared via the solution polymerisation method of the partially neutralised acrylic acid (AA) in the presence of Polyethylene glycol diacrylate (PEGDA), which acted as a cross-linking agent. The SAP post treatment with surface cross-linking showed greater absorbency [39].

4.3 Suspension Polymerisation

The suspension polymerisation is a process to manufacture spherical shape SAP micro particles having a size range of 1 μm to 1 mm. It is also known as inverse suspension polymerisation (ISP). The monomer solution is disseminated in the non-solvent during suspension polymerisation, generating tiny monomer droplets that are stabilised by the addition of a stabiliser. Radicals produced during the heat breakdown of an initiator start the polymerisation process. The unreacted monomers, cross-linking agent, and initiator are subsequently rinse off of the freshly created microparticles. This technique has been used to create certain SAPs microparticles made of poly(hydroxy ethyl methacrylate). Because it makes it simple to remove and handle the potentially dangerous acrylamide monomer that remains in the polymer, the inverse suspension technique has recently become popular for polyacrylamide-based SAPs. Zakaria et al. have synthesised superabsorbent carbonaceous kenaf fibre filled polymer which is known as the superabsorbent carbonaceous fibre polymer (SPC) using inverse suspension polymerisation. The highest water absorbency percentage was found to be with the SPC sample having 0.04 wt% kenaf fibre which was the highest amongst all the samples. It was concluded from their paper that the kenaf fibre filler enhanced the water absorbing capacity of the SAP [40]. In another experiment by Hussain et al. [41] cross-linked poly acrylic acid was prepared by the suspension polymerisation method of the acrylic acid in CO_2. The prepared polymer shows potential increase in the water absorbing capacity and particle size. The surfactants used here act as the stabiliser and helped in producing the desired particle size. When we decrease the amount of acrylic acid disintegrated in the CO_2; it resulted in the production of a smoother surface. A study by Bajpai et al. [42] was done for the preparation of poly methacylic acid superabsorbent hydrogels with the help of ISP method. The prepared SAP by the ISP of methacyclic acid depends on the amount if cross-linking agents and stabilisers during the manufacturing process. The intake of water molecules also depends upon the mixing rate. The activation energy of the swelling process was found to be −23.387 kJ/mol.

5 Physical Synthesis Methods

The physical synthesis method refers to the molecular assembly crosslink by the hydrogen bond, the ionic bond, or the interaction between the polymers. In contrast to the procedure undergoing at room temperature, cryogenic treatment was used to create the cellulose-based superabsorbent hydrogels. This method of preparing the superabsorbent hydrogel is also known as cryogel and is created by the association of strong hydrogen bonds. This potent hydrogen bond may be created during one of the phases of the freeze/thaw cycle, such as when the initial system is frozen or when the sample is stored [43]. Guan et al. have prepared hydrogels with the help of freeze/thaw method, a physical synthesis method, by mixing PVA and chitin. The chitins were mixed homogeneously in the PVA/hemicellulose matrix which helps in restricting the packaging and leads to the enhancement in the thermal stability, crystallinity and compressive strength of the hydrogels. The increment in the freezing cycles leads to more stiffness, stable thermal property and higher crystalline degree [44]. A report about the effect of mechanically induced fragmentation of the polyacrylic superabsorbent polymer (SAP) hydrogel on the properties of cement composites was given by Kalinowski and his co-workers. The reduction in the size of particle, there is increment in the surface area of SAP upto 10 times and when it homogeneously dispersed with the cement matrix, it resulted in greater compatibility with the matrix. When the SAP is dosed in the hydrogel form, it allows improvising in freeze/thaw resistance [45]. Varshney et al. have prepared PVA and soy protein isolate (SPI) based scaffolds that were prepared by using the physical cross-linking method i.e. freeze/thaw method. The water vapour transmission rate (WVTR) ranges from 2000 to 2500 gm^{-2} day^{-1}; which indicates better capacity for wound dressing. Also the scaffolds showed greater stability, mechanical strength, porosity, biocompatibility, degradation rate etc. [46]. According to Wu and his co-workers the integration of cellulose/PVA resulted in maintaining the compressive strength of the concrete under the condition of frost action. Cellulose/PVA hydrogel is initiated into the concrete for improvising the freeze/thaw resistance. The cellulose/PVA hydrogel showed elasticity, greater water retention capacity [47]. Polyaniline/Poly(acrylamide-co-sodium acrylate) porous conductive hydrogels (PASH) were prepared using the freeze/thaw method with excellent mechanical and electrochemical properties. The electrical conductivity is found to be 4.05 S/m along with a tensile strength of 0.12 MPa and elongation break of upto 1245%. Due to the freeze/thaw process, stability also got increased. The PASH-3 samples were found to have the highest specific capacitance value of 849.1 F cm^{-2}. After 1000 charge–discharge cycles, the retention capacitance value also reaches 89% [48]. Electron beam radiation (EBI) method [49] is also one of the physical synthesis method used for the preparation of SAP hydrogels. Preparation and characterisation of the eggshell membrane/PVA hydrogel via electron beam irradiation technique was done by Choi et al. due to the introduction of hydrogel, where it made the network structure more porous and the absorption capacity also increases [50].

6 Use of Synthetic SAPs

Superabsorbent polymers (SAPs) are a class of polymers that are substantially more able to absorb liquids or water than absorbent polymers, often up to several times their own weight. Superabsorbent materials sometimes referred to as hydrogel, absorbent gels, water gels, or supersoakers, are three-dimensional networks of cross linked polymers that are insoluble in water. Super slurpers are cross-linked 3-D networks that include a lot of hydrogel bonds between the solvent and polymeric chains. These bonds serve to maintain the network elastic and stabilise it. Superabsorbent polymer (SAP) materials are hydrophilic networks with enormous water or aqueous solution absorption and retention capacities. They have a 100,000% water uptake capacity [51]. Common SAPs are typically hygroscopic substances that resemble white sugar and are utilised in a variety of applications, including agriculture. The background, kinds, chemical structures, physical and chemical characteristics, testing procedures, applications, and applied research efforts of SAP are all reviewed in this article. Numerous SAP kinds can be created due to the diversity of the potential monomers and macromolecular structure. The two main groups of SAPs—synthetic (petrochemical-based) and natural—were first separated apart (e.g., polysaccharide- and polypeptide-based). However, the majority of the present super absorptions are typically synthesised by solution or inverse-suspension polymerisation from acrylic acid (AA), its salts, and acrylamide (AM). There is a brief description of the primary synthetic (internal) and environmental (external) elements influencing the acrylic anionic SAP properties. The techniques for measuring the practical properties of SAP, including as absorption capacity (both load-free and under load), swelling rate, swollen gel strength, wicking capacity, sol fraction, residual monomer, and ionic sensitivity, were covered [52, 53]. The SAP applications and associated research, in particular in the agricultural and sanitary fields, are examined. In the meantime, the study findings about the impacts of SAP on soil and agricultural successes in Iran, a country with dry conditions, are also covered. Finally, the safety and environmental concerns related to real-world SAP implementations are also covered [54].

7 Natural Polysaccharide Based SAPs

Among all natural superabsorbents, polysaccharides have received the greatest attention due to their unique characteristics and wide range of uses. The two polysaccharides that are most prevalent in nature are cellulose and chitin, which are derived from terrestrial plants and animals, respectively [55]. However, polysaccharides are found in all living things, including seaweed and bacteria [56]. Polysaccharides are organic compounds comprised of a single monosaccharidic species (homopolysaccharides) or a number of monosaccharidic species (heteropolysaccharides) connected together by glycosidic bonds. Alginate, agarose, or chitosan are heteropolysaccharides, while starch, cellulose, or carrageenan are homopolysaccharides [57].

They may also be branching, like dextran or Arabic gum, or linear, like hyaluronic acid. Finally, some of them are non-ionic like guar gum, cationic like chitosan when it dissolves in acidic liquids, and anionic like alginate and carrageenan [58]. The majority of polysaccharides are renewable, harmless, and biodegradable. Additionally, they have a high anisotropy to water due to their charged and hydroxyl groups, which gives them considerable potential as superabsorbent materials. Alginate, chitin and chitosan, starch, carrageenan, agar and agarose, pectin, hyaluronic acid, dextran, cellulose, and gums are the absorbent polysaccharides that are most frequently discussed in the literature. Starch and heteropolysaccharides are used for superabsorbent applications the most [59]. Because chemical alterations can enhance their physicochemical characteristics and provide them new beneficial features, their derivatives are also investigated for superabsorbent applications [58]. To create superabsorbents, for instance, various polysaccharides can be changed by the addition of vinyl groups (such as methacrylate ones). An example of this is that whereas unmodified pectin-based superabsorbents did not exhibit this feature, modified pectin-based superabsorbents demonstrated strong water super absorption in the presence of salt solution [60]. Another type of modified polysaccharide is starch sulphate. Because sulphate groups ionise more readily than carboxylic ones, the hydrogel in question has better salt absorption. In salt solution, there is thus less interaction with mobile ions, and the ionic strength has less of an impact on the swelling of the superabsorbent hydrogel [61].

8 Life Cycle Assessment Method

The life cycle assessment (LCA) method is now often used to evaluate how items and processes affect the environment. Companies are also required to notify customers about product stewardship. However, conducting life cycle inventory (LCI) and gathering data on product stewardship are pricey. ERP (enterprise resource planning) systems are frequently used in enterprises to plan, organise, direct, measure, and control their financial, human, and logistical resources. ERP systems are further equipped to record environmental management data, such as raw material and supply weights, energy and water usage. Bills of materials and work orders, among other information necessary for doing life cycle inventories, are included in ERP systems.

Januschkowetz et al. explain in their article how the ERP system SAP R/3 may be used to construct an LCI on a product. Additionally, with the aid of an ERP system, supply allocation techniques at the product level are illustrated. The paper provides an illustration of how an LCI of an electric product from an automotive supplier was carried out utilising the SAP R/3 system [62].

As a result of eco-innovation methods aimed at lowering the use of non-renewable raw resources, biobased goods are being created, produced, and employed in sustainable production and consumption strategies. However, life-cycle based research must be conducted in order to fully assess the environmental profile of biobased goods, since simply being biobased does not equate to being regarded ecologically beneficial. The term "bio-based" refers to goods created from renewable biotic raw resources like plants and trees, hence bio-based goods might take the place of those manufactured from fossil fuels [63]. This definition covers a wide range of potential applications across a number of market categories, including those for bioplastics and biofuels that are experiencing rapid expansion and substantial diversification [64]. Different forms of bioplastics, including starch-based plastics like polylactic acid (PLA) and Mater-bi® and bio-based or partially bio-based commodity plastics like bio-based polyethylene and polyethylene terephthalate, are processed in these products depending on the application. The necessity to develop alternatives to fossil fuels and petrochemical compounds has resulted in a significant market increase for both biofuels and bioplastics in recent years. The market is expected to rise by more than 1.7 million tonnes by the year 2015, at a rate of almost 20% annually. The capacity of bio-based plastics might reach 2.3 Mt globally by 2013 and 3.5 Mt globally by 2020 [65]. Other predictions indicate that manufacturing of bio-based goods will expand in the near future, thus it's critical to comprehend the potential and risks to these resources [66]. The development of new disposal choices, possible cost savings, and enhanced technological features all support the market expansion. Due to rising prices, increased environmental awareness regarding climate change, and a rigid reliance on fossil fuels, bioplastics also enjoy strong market approval. Despite these benefits, a number of academics, global movements, and nongovernmental organisations (NGOs) for nature conservation and social equity raise a number of issues regarding the sustainability of bio-based products, highlighting the possibility of land competition. Although research is being done to employ lignocelluloses as a feedstock, practically all commercially accessible bio-based goods are now manufactured from crops that are grown for their starch or sugar [67]. The sustainability of bio-based products must thus be assessed using an integrated framework that takes into account issues related to food, the environment, and energy as well as enhancing land productivity rather than enlarging already-cultivated areas [68]. Furthermore, according to Shen et al., there is not enough sustainably generated biomass to meet all of the demands for bioresources (including food, fuel, and materials), thus food production must take precedence [65]. The argument over science is open and contentious.

8.1 Disposal Scenarios

Systems for managing waste and energy are changing in many nations. The threat of global climate change, as well as other factors including ozone depletion, acidification, toxicity, resource consumption, and depletion, is the main environmental factor driving the changes. It is crucial to take the environment into account when making new strategic decisions on waste and energy management systems. When developing trash policies, a hierarchy for waste management is frequently proposed and applied [69]. Because it comprises several concepts and procedures, waste management is a complicated process. These include practices and tools for creating, maintaining, storing, gathering, transferring, moving, processing, and disposing of garbage [70]. All of these procedures must adhere to current social and legal norms, safeguard the environment and public health, and be acceptable in terms of both aesthetics and economics [71]. The wastes produced by the oil and gas industries are extremely diverse in their properties and huge in quantity, with many of them being dangerous in nature [72]. The environmental issues in the oil and gas industry are a result of poor judgements. New management approaches should be used, and waste management systems should be assessed, in order to ensure a sustainable development. By identifying and measuring the energy and materials required, as well as the waste discharged into the environment, life-cycle assessment (LCA) is best understood as an objective technique for evaluating the environmental impacts associated with a product, process, or activity. In life-cycle assessment, possibilities to enable environmental improvements are assessed and put into action. In other words, LCA considers the difficulties that other environmental management techniques, such as environmental performance evaluation, environmental auditing, material, energy, and toxic-analysis, etc., do not take into consideration. Unlike other pollution management strategies that place emphasis on one of the aforementioned challenges, such as recovery and toxicity reduction, LCA may take a variety of factors into account. On beverage containers, the first life cycle study was performed in 1969. Finding the container type that had the least impact on the environment and natural resources was the main goal of the investigation. Identification of energy and material flows was the end outcome, however the environmental effect was not assessed [73]. Two disposal techniques—land filling without energy recovery and incineration with energy recovery—were examined from the perspectives of energy consumption and greenhouse gas emissions in the research of MSW management in Phuket, a province of Thailand [74]. In a different research, LCA was used to compare the current management system with the system that was suggested for the future in order to assess the environmental effects of managing the fermentable portion of garbage in the Barcelona Metropolitan Area (BMA) [75]. In order to establish a technology with a broad perspective and in a rigorous and objective manner to allow water reuse to that purified waters, LCA was implemented in a study on global environmental analysis of waste water treatment and some possible additional tertiary treatments allowing water reuse to that purified waters [76]. The environmental effects of various industrial waste disposal methods, sulphur waste, and used oil scenarios have all been looked into and modelled in this study. The methodologies have been explored in further detail using a real situation at an oil refinery as a case study, with use of LCA shown to help in the production of alternatives and to offer the decision maker with insightful information [77]. Among the techniques for evaluating environmental impacts, including economic input–output analysis, risk analysis, strategic impact analysis, etc., the LCA approach was chosen. Environmental concerns and burdens were quantified using this methodology to make comparisons easier. The solutions with the fewest negative effects were ultimately recommended. This study was conducted in connection with the LCA of the disposal of sulphur wastes and used lubricating oils in Tehran, Iran, during 2011.

8.2 Standard MSW Collection

Waste management is forced to pursue more efficient collection plans that are technically feasible, ecologically effective, and economically sustainable due to the rise in municipal solid waste (MSW) production over the past few years. For the purpose of raising service quality, performance metrics must be used to evaluate MSW services [78]. In this study, we emphasise the value of routine system monitoring as a tool for evaluating services. To highlight the strengths and weaknesses of the collection system and to aid proactive management decision-making and strategic planning, we specifically choose and test a core set of MSW collection performance indicators (effective collection distance, effective collection time, and effective fuel consumption). Data from the Oporto Municipality in Portugal were acquired using a mixed collecting technique over the course of a year, one week every month, and were used in a statistical analysis. This research offers a functional assessment of the collection circuits and enables efficient short-term municipal collection plans at the level of, for instance, collection frequency and schedules, and container type. Wastes produced by our society's needs have a tendency to grow in amount as living standards rise. Reducing the quantity of garbage produced is the most efficient way to solve the issue [79]. This is in contrast to the trend of steadily rising living standards. Due to the depletion of inexpensive land resources and the wastefulness of disposing of valuable resources during the landfill operation, the chances for land filling as a way of disposing of municipal solid waste (MSW) are fast dwindling. Separation, recycling, and resource recovery are available alternatives for reducing waste in the public sector. In the past 20 years, resource recovery in the form of producing heat and electricity has gained popularity due to the limited economic benefits of separation and recycling [80]. Although MSW incineration has had some unrest in terms of popularity throughout this time, it is still a desirable disposal option and offers a number of advantages.

- The volume and mass of MSW are diminished by 85–90% by volume, to a small portion of their original size.
- The decrease in waste is instantaneous and not reliant on slow biological reaction periods.
- To cut down on transportation expenses, incineration plants might be built nearer to MSW sources or collecting sites.
- Energy sales can help defray the cost of the operation, and air emissions can be regulated to stay under legal limits for the environment.

8.3 Recycling, No SAP Recovery

This study's goal is to assess how recovering superabsorbent polymers from absorbent hygiene products affects the life cycle (AHPs). AHPs, which include adult incontinence pads, feminine hygiene products, and diapers for infants, have a significant negative influence on the environment. The bulk of these items, which frequently include mixtures of polypropylene, polyethylene, elastics, cellulose, and superabsorbent polymers (SAPs), are single-use despite being practical. The way AHPs are now disposed of causes the production of significant amounts of municipal solid waste, the loss of precious materials like SAPs, and higher manufacturing costs. Despite growing awareness of AHP consequences, it's vital to remember that resource extraction and manufacturing—rather than disposal—resulting in the greatest AHP life cycle impacts. Due to their significant upstream life cycle contributions, the SAPs in these goods are particularly significant. By concentrating on the possibility for SAP recovery and re-use, we want to give insight on how we may decrease upstream effects. The life cycle method is used in this study to assess the alternatives to the traditional disposal of AHPs for infant diapers in Europe. We consider the following possibilities when we examine the environmental trade-offs: (1) dumping or burning used diapers; (2) recycling used diapers without potential SAP recovery; and (3) recycling used diapers. There are five recycling scenarios in all, and each one is examined under two allocation criteria. The SimaPro LCA programme was used to model environmental effects using the ReCipe 2016 impact assessment framework. The study's findings indicate that SAP recovery has the ability to reduce life cycle emissions by 35% when compared to recycling methods and by 54% when compared to regular landfilling and incineration. SAP reuse and recovery also has the potential to significantly reduce the energy and water costs associated with SAP manufacture.

We intend to make clear the point at which SAP recovery indicates the potential for a circular economy by evaluating these environmental effects [81].

Pre-consumer kid and adult AHPs are composed of a blended absorbent core, back sheet, acquisition and distribution layer, and porous top sheet. Although other component configurations also include adhesives, elastic, dyes, and lotions, these layers typically contain about 43–59% cellulose, 14–27% superabsorbent polymer (SAP), which is frequently sodium polyacrylate, and 19–22% polyolefins (polypropylene, polyethylene) [82–84]. After consumer usage, the amounts of cellulose, polyolefins, and SAP in incontinence AHPs are around 27%, 12%, and 5%, respectively, but the contents of the faeces and urine range from about 4–6% and >50%, respectively [85]. Although the raw materials used to produce AHPs are of extremely good quality [82], the purpose of this study is to determine if post-consumer AHPs have sufficient value to warrant further recycling [7]. Many mineral tailings have a high process water content, which calls for the development and use of innovative, efficient dewatering techniques that use high-capacity, regenerable water-absorbing polymers. Superabsorbent (SAB) polymers' stimulus responsiveness enables them to release absorbed water in response to changing process conditions (such as temperature and pH), allowing superabsorbent polymers to be recycled to increase the economics of the dewatering process. This study examines recovery of the absorbed water by pH-induced superabsorbent regeneration in order to comprehend the impact of process parameters and ore type coupled with cost effectiveness of the superabsorbent dewatering system. Prior to the experiments, the superabsorbent, sodium polyacrylate, absorbed water from hydrophilic saprolitic laterite and hydrophobic chalcopyrite slurries. In particular, the impact of absorbed water pH, SAB dosage, and acid type and dosage (CH_3COOH, HCl) on the SAB's water recovery behaviour was investigated. The highest water recovery (95%) from both slurries was made possible by the SAB dosage of 2 g/100 g slurry and the HCl dosage of 2 g/100 g absorbed water, although the pH and mineralogy of the slurries had no discernible impact on the SAB dewatering procedure. In the SAB dewatering system, CH3COOH performed less well overall than HCl in terms of the amount of water that could be recovered, although at slightly acidic solution conditions (pH 4), Ch3COOH recovered substantially more water (about 70%) than HCl (about 45 wt%). The findings contribute to the improvement of hydrophobic and hydrophilic materials handling at a lower cost [82].

8.4 *Recycling + SAP Recovery*

In certain areas, some used paper diapers are recycled, primarily by open-loop recycling or thermal recycling. However, the bulk of used paper diapers are disposed of thorough burning. To date, a number of strategies for recycling old paper diapers have been put out and developed, however these strategies are thought to employ various types and quantities of recovered materials and to perform differently in terms of the environment. The use of recycled pulp and superabsorbent polymer (SAP) as

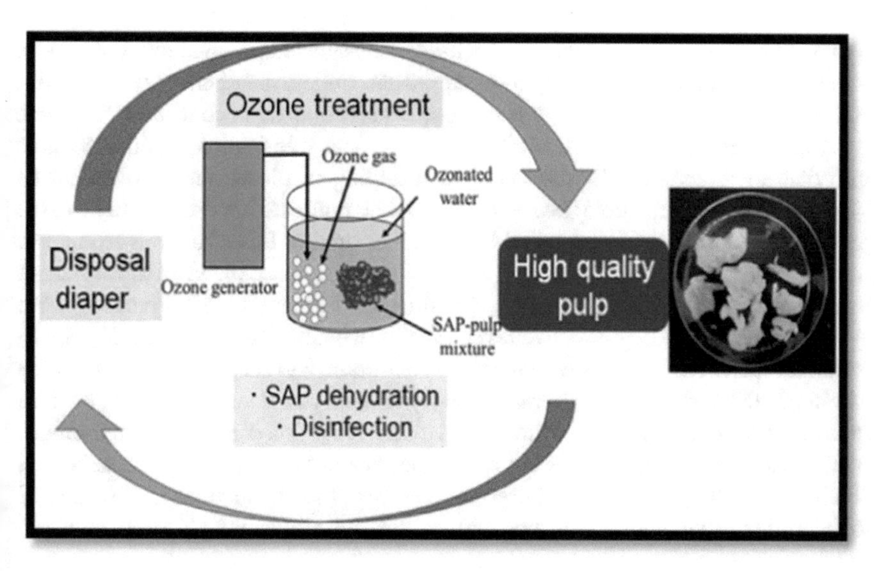

Fig. 3 Schematic illustration for Ozone treatment of disposal diaper [88]

materials for paper diapers was evaluated via the environmental impact using the life cycle assessment (LCA) method, using data obtained from experimental facilities for recycling, in this study. A new technology was developed for the closed-loop recycling of used paper diapers. The results show a greenhouse gas decrease of 47% and 39%, respectively, when the novel technology is compared to the landfill and incinerator operations. The findings also demonstrate that such recycling is anticipated to minimise water use and land occupation, two factors that are directly connected to pulp, the primary raw material used to make paper diapers [86]. The pulp from used diapers has been recycled in a manner. The pulp from the used diapers was recycled after the super-absorbent polymer (SAP) in the diapers was dehydrated using ozone oxidation. After washing and centrifugally separating the diaper waste, the SAP and pulp combination was oxidised with ozone to separate the SAP and pulp and enable the recycling of the pulp. A sample of SAP was diluted in 2000 mL of a 45 g/m^3 ozone solution by adding ozone at a flow rate of 1.0 mL/min after it had absorbed either 400 mL of distilled water or 80 mL of a 0.9% NaCl solution (as shown in Fig. 3). As the reaction time lengthened, the SAP concentration fell. The cross-linked SAP structure was oxidised by ozone, which removed the SAP's capacity to hold water. The SAP's cross-linking agent's ether bonds were likewise broken by ozone treatment. As a result, the water-insoluble SAP was changed into a sodium acrylate polymer. As the reaction time lengthened, the SAP's molecular weight and degree of dispersion likewise reduced. The recovered pulp's water absorption was greater than the recommended limit of 10 g/g and was practically identical to that of virgin pulp. The ozone treatment had no effect on the recovered pulp's water absorptivity. We came to the conclusion that pulp from disposable diapers may be recycled using the procedure [87, 88].

The typical organic material found in used disposable diapers includes fluffy pulp, urine, and or excreta, as well as nonwoven textiles, Super Absorbent Polymer (SAP), and organic material. Currently, this waste stream is being dumped in landfills, which is having a negative impact on the environment. The valorisation of biodegradable materials by anaerobic digestion and the recycling of plastic and SAP might be alternate management strategies. Nappies must be pretreated in order to remove the plastic and SAP from the organic substance. The goal of this effort was to create a method to reduce the amount of SAP, a component that may expand up to 1500 times its own mass due to water absorption, impeding any further biological processes. We evaluated the deswelling effectiveness of $CaCl_2$, $MgCl_2$, and various $CaCl_2/MgCl_2$ combinations on SAP, residual reagent concentration, and reagent cost. The combination of 20% $CaCl_2$ and 50% $MgCl_2$ (w/w) of SAP was determined to be the best salts to use in order to achieve a final SAP volume reduction of 92.7% at a reasonable cost. An appropriate COD: N ratio for further anaerobic digestion processing was obtained after the hydrolysate of used diapers was physicochemically characterised to determine its suitability as a substrate for anaerobic digestion [11, 89].

9 Future Scope

The effectiveness of a superabsorbent polymer may degrade over time when exposed to real-world circumstances. These shifts have only been explored in the lab and during concise periods of time. It is important to consider how these materials will function over the course of several years if they are to be used in the field. Studies of SAP deterioration in a variety of outdoor settings are needed. Benefit-to-cost ratios for various SAPs may be determined by field-scale evaluation of depth, rate, and application technique.

10 Conclusion and Out Look

Superabsorbent polymers are able to rapidly absorb and retain a significant amount of liquid when exposed to water or other aqueous solutions. Their three-dimensional polymeric network structure allows them to greatly expand when in contact with water without dissolving. Synthetic superabsorbent polymers are progressively replacing their natural equivalents because of their better absorbency, accessibility to raw materials, and durability. The primary application of superabsorbent polymers has been to expand the amount of time between irrigations by enhancing the soil's ability to store water. Evidence of concept for SAPs' capacity to lessen compaction has proved promising. However, most SAP research were conducted in the lab or on small field settings, and results may not transfer to larger scale projects or may not be economically feasible at such a scale. Evaluation of the function, performance, and overall advantages to crop yields and plant productivities is best accomplished through large-scale, long-term initiatives.

References

1. Ekebafe, L.O., Ogbeifun, D.E., Okieimen, F.E.: Polymer applications in agriculture. Biokemistri **23**(2) (2011)
2. Zohuriaan-Mehr, M.J., Kabiri, K.: Superabsorbent polymer materials: a review. Iranian Polym. J. **17**, 451–477 (2008)
3. Bai, W., Zhang, H., Liu, B., Wu, Y., Song, J.Q.: Effects of super-absorbent polymers on the physical and chemical properties of soil following different wetting and drying cycles. Soil Use Manag. **26**, 253–260 (2010)
4. Tsigkou, K., Zagklis, D., Vasileiadi, A., Kostagiannakopoulou, C., Sotiriadis, G., Anastopoulos, I., Kornaros, M.: Used disposable nappies: environmental burden or resource for biofuel production and material recovery? Resour. Conserv. Recycl. **185**, 106493 (2022)
5. Hait, A., Powers, S.E.: The value of reusable feminine hygiene products evaluated by comparative environmental life cycle assessment. Resour. Conserv. Recycl. **150**, 104422 (2019)
6. Cordella, M., Bauer, I., Lehmann, A., Schulz, M., Wolf, O.: Evolution of disposable baby diapers in Europe: life cycle assessment of environmental impacts and identification of key areas of improvement. J. Clean. Prod. **95**, 322–331 (2015)
7. Somers, M.J., Alfaro, J.F., Lewis, G.M.: Feasibility of superabsorbent polymer recycling and reuse in disposable absorbent hygiene products. J. Clean. Prod. **313**, 127686 (2021)
8. Vink, E.T., Rabago, K.R., Glassner, D.A., Gruber, P.R.: Applications of life cycle assessment to NatureWorks™ polylactide (PLA) production. Polym. Degrad. Stabil. **80**(3), 403–419 (2003)
9. Saberi, S., Kouhizadeh, M., Sarkis, J., Shen, L.: Blockchain technology and its relationships to sustainable supply chain management. Int. J. Prod. Res. **57**(7), 2117–2135 (2019)
10. Ali, S., Sarhanis, A., Turn, C., McLaughry, E., Hartin, K., Hayes, M.: Sustainability Assessment: Seventh Generation Diapers Versus gDiapers (2011)
11. Khoo, S.C., Phang, X.Y., Ng, C.M., Lim, K.L., Lam, S.S., Ma, N.L.: Recent technologies for treatment and recycling of used disposable baby diapers. Process Saf. Environ. Prot. **123**, 116–129 (2019)
12. Płotka-Wasylka, J., Makoś-Chełstowska, P., Kurowska-Susdorf, A., Treviño, M.J.S., Guzmán, S.Z., Mostafa, H., Cordella, M.: End-of-life management of single-use baby diapers: analysis of technical, health and environment aspects. Sci. Total Environ. 155339 (2022)
13. Mishra, R., Singh, R.K., Rana, N.P.: Developing environmental collaboration among supply chain partners for sustainable consumption & production: insights from an auto sector supply chain. J. Clean. Prod. **338**, 130619 (2022)
14. Kiatkamjornwong, S.: Superabsorbent polymers and superabsorbent polymer composites. ScienceAsia **33**(1), 39–43 (2007)
15. Tripathi, A., Melo, J.S.: An overview on super absorbent polymer and its application in agriculture. Indian Assoc. Nuclear Chem. Allied Sci. **38**
16. Zhang, T., Li, Z., Lü, Y., Liu, Y., Yang, D., Li, Q., Qiu, F.: Recent progress and future prospects of oil-absorbing materials. Chin. J. Chem. Eng. **27**(6), 1282–1295 (2019)
17. Ekebafe, M.O., Ekebafe, L.O., Maliki, M.: Utilisation of biochar and superabsorbent polymers for soil amendment. Sci. Prog. **96**(1), 85–94 (2013)

18. Wang, B., Liang, W., Guo, Z., Liu, W.: Biomimetic super-lyophobic and super-lyophilic materials applied for oil/water separation: a new strategy beyond nature. Chem. Soc. Rev. **44**(1), 336–361 (2015)
19. Ullah, F., Othman, M.B.H., Javed, F., Ahmad, Z., Akil, H.M.: Classification, processing and application of hydrogels: a review. Mater. Sci. Eng. C **57**, 414–433 (2015)
20. Behera, S., Mahanwar, P.A.: Superabsorbent polymers in agriculture and other applications: a review. Polym.-Plast. Technol. Mater. **59**(4), 341–356 (2020)
21. Venkatachalam, D., Kaliappa, S.: Superabsorbent polymers: a state-of-art review on their classification, synthesis, physicochemical properties, and applications. Rev. Chem. Eng. (2021)
22. Romasn-Hess, A.Y., Feldkamp, J.R.: U.S. Patent No. 6,855,434. U.S. Patent and Trademark Office, Washington, DC (2005)
23. Shukla, N.B., Rattan, S., Madras, G.: Swelling and dye-adsorption characteristics of an amphoteric superabsorbent polymer. Ind. Eng. Chem. Res. **51**(46), 14941–14948 (2012)
24. Rehman, T.U., Shah, L.A., Khan, M., Irfan, M., Khattak, N.S.: Zwitterionic superabsorbent polymer hydrogels for efficient and selective removal of organic dyes. RSC Adv. **9**(32), 18565–18577 (2019)
25. Lalani, R., Liu, L.: Electrospun zwitterionic poly (sulfobetaine methacrylate) for nonadherent, superabsorbent, and antimicrobial wound dressing applications. Biomacromolecules **13**(6), 1853–1863 (2012)
26. Masuda, F.: Superabsorbent Polymers, p. 6. Japan Polymer Society, Kyoritsu Shuppann (1987)
27. Xu, T., Zhu, W., Sun, J.: Structural modifications of sodium polyacrylate-polyacrylamide to enhance its water absorption rate. Coatings **12**(9), 1234 (2022)
28. Ghasri, M., Bouhendi, H., Kabiri, K., Zohuriaan-Mehr, M.J., Karami, Z., Omidian, H.: Superabsorbent polymers achieved by surface cross linking of poly (sodium acrylate) using microwave method. Iranian Polym. J. **28**(7), 539–548 (2019)
29. Oyama, Y., Hiejima, Y., Nitta, K.H.: Swelling behavior of butyl and chloroprene rubber composites with poly (sodium acrylate) showing high water uptake. J. Appl. Polym. Sci. **137**(14), 48535 (2020)
30. Mehra, S., Nisar, S., Chauhan, S., Singh, G., Singh, V., Rattan, S.: A dual stimuli responsive natural polymer based superabsorbent hydrogel engineered through a novel cross-linker. Polym. Chem. **12**(16), 2404–2420 (2021)
31. Gök, M.K., Demir, K., Cevher, E., Özsoy, Y., Cirit, Ü., Bacınoğlu, S., Pabuccuoğlu, S.: The effects of the thiolation with thioglycolic acid and l-cysteine on the mucoadhesion properties of the starch-graft-poly (acrylic acid). Carbohydr. Polym. **163**, 129–136 (2017)
32. Relleve, L.S., Aranilla, C.T., Barba, B.J.D., Gallardo, A.K.R., Cruz, V.R.C., Ledesma, C.R.M., Abad, L.V.: Radiation-synthesized polysaccharides/polyacrylate super water absorbents and their biodegradabilities. Radiat. Phys. Chem. **170**, 108618 (2020)
33. Bahadoran Baghbadorani, N., Behzad, T., Karimi Darvanjooghi, M.H., Etesami, N.: Modelling of water absorption kinetics and biocompatibility study of synthesized cellulose nanofiber-assisted starch-graft-poly(acrylic acid) hydrogel nanocomposites. Cellulose **27**(17), 9927–9945 (2020)
34. Mohite, P.B., Adhav, S.S.: A hydrogels: methods of preparation and applications. Int. J. Adv. Pharm. **6**(3), 79–85 (2017)
35. Liu, Z.S., Rempel, G.L.: Preparation of superabsorbent polymers by crosslinking acrylic acid and acrylamide copolymers. J. Appl. Polym. Sci. **64**(7), 1345–1353 (1997)
36. Bhattacharya, S.S., Mishra, A., Pal, D., Ghosh, A.K., Ghosh, A., Banerjee, S., Sen, K.K.: Synthesis and characterization of poly (acrylic acid)/poly (vinyl alcohol)-xanthan gum interpenetrating network (IPN) superabsorbent polymeric composites. Polym. Plast. Technol. Eng. **51**(9), 878–884 (2012)
37. Tan, Y., Chen, H., Chen, M.: Preparation, characterization and hydration-retarding mechanism of a new type of highly alkali-resistant superabsorbent polymer for sustainable concrete structures. Constr. Build. Mater. **367**, 130263 (2023)
38. Ma, X., Wen, G.: Development history and synthesis of super-absorbent polymers: a review. J. Polym. Res. **27**(6), 1–12 (2020)

39. Chen, Z., Liu, M., Ma, S.: Synthesis and modification of salt-resistant superabsorbent polymers. React. Funct. Polym. **62**(1), 85–92 (2005)
40. Moini, N., Kabiri, K., Zohuriaan-Mehr, M.J.: Practical improvement of SAP hydrogel properties via facile tunable cross-linking of the particles surface. Polym. Plast. Technol. Eng. **55**(3), 278–290 (2016)
41. Zakaria, M.E.B.T., Jamari, S.S.B., Ghazali, S.: Synthesis of superabsorbent carbonaceous kenaf fibre filled polymer via inverse suspension polymerisation. J. Mech. Eng. Sci. **11**(3), 2794 (2017)
42. Hussain, Y.A., Liu, T., Roberts, G.W.: Synthesis of cross-linked, partially neutralized poly (acrylic acid) by suspension polymerization in supercritical carbon dioxide. Ind. Eng. Chem. Res. **51**(35), 11401–11408 (2012)
43. Bajpai, S.K., Bajpai, M., Sharma, L.: Inverse suspension polymerization of poly (methacrylic acid-co-partially neutralized acrylic acid) superabsorbent hydrogels: synthesis and water uptake behavior. Des. Monomers Polym. **10**(2), 181–192 (2007)
44. Ma, J., Li, X., Bao, Y.: Advances in cellulose-based superabsorbent hydrogels. RSC Adv. **5**(73), 59745–59757 (2015)
45. Guan, Y., Bian, J., Peng, F., Zhang, X.M., Sun, R.C.: High strength of hemicelluloses based hydrogels by freeze/thaw technique. Carbohydr. Polym. **101**, 272–280 (2014)
46. Kalinowski, M., Woyciechowski, P., Sokołowska, J.: Effect of mechanically-induced fragmentation of polyacrylic superabsorbent polymer (SAP) hydrogel on the properties of cement composites. Constr. Build. Mater. **263**, 120135 (2020)
47. Varshney, N., Sahi, A.K., Poddar, S., Vishwakarma, N.K., Kavimandan, G., Prakash, A., Mahto, S.K.: Freeze–thaw-induced physically cross-linked superabsorbent polyvinyl alcohol/ soy protein isolate hydrogels for skin wound dressing: in vitro and in vivo characterization. ACS Appl. Mater. Interfaces **14**(12), 14033–14048 (2022)
48. Wu, K., Han, H., Xu, L., Gao, Y., Yang, Z., Jiang, Z., De Schutter, G.: The improvement of freezing–thawing resistance of concrete by cellulose/polyvinyl alcohol hydrogel. Constr. Build. Mater. **291**, 123274 (2021)
49. Chen, L., Zhang, W., Dong, Y., Chen, Q., Ouyang, W., Li, X., Huang, J.: Polyaniline/poly (acrylamide-co-sodium acrylate) porous conductive hydrogels with high stretchability by freeze-thaw-shrink treatment for flexible electrodes. Macromol. Mater. Eng. **305**(3), 1900737 (2020)
50. Choi, J., Pant, B., Lee, C., Park, M., Park, S.J., Kim, H.Y.: Preparation and characterization of eggshell membrane/PVA hydrogel via electron beam irradiation technique. J. Ind. Eng. Chem. **47**, 41–45 (2017)
51. Ahmed, E.M.: Hydrogel: preparation, characterization, and applications: a review. J. Adv. Res. **6**(2), 105–121 (2015)
52. Mignon, A., De Belie, N., Dubruel, P., Van Vlierberghe, S.: Superabsorbent polymers: a review on the characteristics and applications of synthetic, polysaccharide-based, semi-synthetic and 'smart' derivatives. Eur. Polym. J. **117**, 165–178 (2019)
53. Van den Heede, P., Mignon, A., Habert, G., De Belie, N.: Cradle-to-gate life cycle assessment of self-healing engineered cementitious composite with in-house developed (semi-) synthetic superabsorbent polymers. Cem. Concr. Compos. **94**, 166–180 (2018)
54. Zohourian, M.M., Kabiri, K.: Superabsorbent polymer materials: a review (2008)
55. Kabiri, K., Omidian, H., Zohuriaan-Mehr, M.J., Doroudiani, S.: Superabsorbent hydrogel composites and nanocomposites: a review. Polym. Compos. **32**(2), 277–289 (2011)
56. Salmon, S., Hudson, S.M.: Crystal morphology, biosynthesis, and physical assembly of cellulose, chitin, and chitosan. J. Macromol. Sci. Part C: Polym. Rev. **37**(2), 199–276 (1997)
57. Beaumont, M., Tran, R., Vera, G., Niedrist, D., Rousset, A., Pierre, R., Forget, A.: Hydrogel-forming algae polysaccharides: from seaweed to biomedical applications. Biomacromolecules **22**(3), 1027–1052 (2021)
58. Llanes, L., Dubessay, P., Pierre, G., Delattre, C., Michaud, P.: Biosourced polysaccharide-based superabsorbents. Polysaccharides **1**(1), 51–79 (2020)

59. Mittal, H., Ray, S.S., Okamoto, M.: Recent progress on the design and applications of polysaccharide-based graft copolymer hydrogels as adsorbents for wastewater purification. Macromol. Mater. Eng. **301**(5), 496–522 (2016)
60. Kaczmarek, B., Nadolna, K., Owczarek, A.: The physical and chemical properties of hydrogels based on natural polymers. In: Hydrogels Based on Natural Polymers, pp. 151–172 (2020)
61. Guilherme, M.R., Aouada, F.A., Fajardo, A.R., Martins, A.F., Paulino, A.T., Davi, M.F., Muniz, E.C.: Superabsorbent hydrogels based on polysaccharides for application in agriculture as soil conditioner and nutrient carrier: a review. Eur. Polym. J. **72**, 365–385 (2015)
62. Raafat, A.I., Eid, M., El-Arnaouty, M.B.: Radiation synthesis of superabsorbent CMC based hydrogels for agriculture applications. Nuclear Instrum. Methods Phys. Res. Sect. B: Beam Interact. Mater. Atoms **283**, 71–76 (2012)
63. Januschkowetz, A.: Use of Enterprise Resource Planning Systems for Life Cycle Assessment and Product Stewardship. Carnegie Mellon University (2001)
64. Arikan, E.B., Ozsoy, H.D.: A review: investigation of bioplastics. J. Civ. Eng. Archit. **9**(2), 188–192 (2015)
65. Calabrò, P.S., Grosso, M.: Bioplastics and waste management. Waste Manag. **78**, 800–801 (2018)
66. Shen, Y., Delaglio, F., Cornilescu, G., Bax, A.: TALOS+: a hybrid method for predicting protein backbone torsion angles from NMR chemical shifts. J. Biomol. NMR **44**(4), 213–223 (2009)
67. Kemp-Benedict, E.: Telling better stories: strengthening the story in story and simulation. Environ. Res. Lett. **7**(4), 041004 (2012)
68. Chen, G.Q., Patel, M.K.: Plastics derived from biological sources: present and future: a technical and environmental review. Chem. Rev. **112**(4), 2082–2099 (2012)
69. Harvey, M., Pilgrim, S.: The new competition for land: food, energy, and climate change. Food Policy **36**, S40–S51 (2011)
70. Finnveden, G.: On the limitations of life cycle assessment and environmental systems analysis tools in general. Int. J. Life Cycle Assess. **5**(4), 229–238 (2000)
71. Nouri, A., Etminan, A., Teixeira da Silva, J.A., Mohammadi, R.: Assessment of yield, yield-related traits and drought tolerance of durum wheat genotypes (Triticum turjidum var. durum Desf.). Aust. J. Crop Sci. **5**(1), 8–16 (2011)
72. Nouri, J., Nouri, N., Moeeni, M.: Development of industrial waste disposal scenarios using life-cycle assessment approach. Int. J. Environ. Sci. Technol. **9**(3), 417–424 (2012)
73. Elshorbagy, W., Alkamali, A.: Solid waste generation from oil and gas industries in United Arab Emirates. J. Hazard. Mater. **120**(1–3), 89–99 (2005)
74. LeVan, S.L.: Life cycle assessment: measuring environmental impact. In: 49th Annual Meeting of the Forest Prodcust Society (1995)
75. Liamsanguan, C., Gheewala, S.H.: LCA: a decision support tool for environmental assessment of MSW management systems. J. Environ. Manag. **87**(1), 132–138 (2008)
76. Güereca, L.P., Gassó, S., Baldasano, J.M., Jiménez-Guerrero, P.: Life cycle assessment of two biowaste management systems for Barcelona, Spain. Resour. Conserv. Recycl. **49**(1), 32–48 (2006)
77. Ortiz, M., Raluy, R.G., Serra, L.: Life cycle assessment of water treatment technologies: wastewater and water-reuse in a small town. Desalination **204**(1–3), 121–131 (2007)
78. Weston, N., Clift, R., Holmes, P., Basson, L., White, N.: Streamlined life cycle approaches for use at oil refineries and other large industrial facilities. Ind. Eng. Chem. Res. **50**(3), 1624–1636 (2011)
79. Teixeira, C.A., Avelino, C., Ferreira, F., Bentes, I.: Statistical analysis in MSW collection performance assessment. Waste Manag. **34**(9), 1584–1594 (2014)
80. Tai, J., Zhang, W., Che, Y., Feng, D.: Municipal solid waste source-separated collection in China: a comparative analysis. Waste Manag. **31**(8), 1673–1682 (2011)
81. Singhal, S., Pandey, S.: Solid waste management in India: status and future directions. TERI Inf. Monit. Environ. Sci. **6**(1), 1–4 (2001)
82. Takaya, C.A., Cooper, I., Berg, M., Carpenter, J., Muir, R., Brittle, S., Sarker, D.K.: Offensive waste valorisation in the UK: assessment of the potentials for absorbent hygiene product (AHP) recycling. Waste Manag. **88**, 56–70 (2019)

83. Das, D.: Composite nonwovens in absorbent hygiene products. Compos. Non-Woven. Mater. 74–88 (2014)
84. Kim, K.S., Cho, H.S.: Pilot trial on separation conditions for diaper recycling. Waste Manag. **67**, 11–19 (2017)
85. Deloitte, U.K.: Absorbent hygiene products comparative life cycle assessment (2011)
86. Joseph-Soly, S., Asamoah, R., Addai-Mensah, J.: Superabsorbent recycling for process water recovery. Chem. Eng. J. Adv. **6**, 100085 (2021)
87. Itsubo, N., Wada, M., Imai, S., Myoga, A., Makino, N., Shobatake, K.: Life cycle assessment of the closed-loop recycling of used disposable diapers. Resources **9**(3), 34 (2020)
88. Ichiura, H., Nakaoka, H., Konishi, T.: Recycling disposable diaper waste pulp after dehydrating the superabsorbent polymer through oxidation using ozone. J. Clean. Prod. **276**, 123350 (2020)
89. Tsigkou, K., Tsafrakidou, P., Zafiri, C., Beobide, A.S., Kornaros, M.: Pretreatment of used disposable nappies: super absorbent polymer deswelling. Waste Manag. **112**, 20–29 (2020)

Bio-based Superabsorbent Polymer: Current Trends, Applications and Future Scope

Roshni Pattanayak and Tapaswini Jena

Abstract Super absorbent polymers (SAPs) are three-dimensional polymeric network structured macromolecules that can able to absorb liquid substances such as water, saline solution, body fluid, etc. in large amounts to their original mass. These show the characteristics of water insoluble and swelling behaviour in an aqueous medium for which these are utilised in various fields, especially in agricultural, health and hygiene applications. Polyacrylate-based synthetic SAPs are nowadays widely used materials as an absorbent because of their significant absorbing properties. But its adverse effect on both health and the environment brings a matter of concern among researchers to think about natural-based superabsorbent. Thus, bio-based superabsorbent materials are a green effective alternative to synthetic ones, which can be utilised in various fields without compromising on their properties. In this context, we focused on the sources, current trends, application and future scope of bio-based SAPs which would help the researcher by collecting brief ideas on it.

1 Introduction

Superabsorbent polymers are polymeric macromolecules that comprised a large amount of hydroxyl group (–OH group) to form a three-dimensional network structure. For the presence of this hydrophilic group, SAPs can able to absorb the aqueous fluid such as water, saline solution, body fluid, etc. in large amounts to their original mass to form the hydrogel crosslinked structure. This hydrogel is a water-insoluble material that can able to swell in absorbing water and also retain the liquid in its structure which makes SAPs an ideal material as an absorbent. Generally, Super absorbent polymers are classified into two categories, i.e., Natural-based and Petrochemical-based SAPs. These SAPs are formed by crosslinked polymer with covalently bonded structures categorised into three types such as first-generation, second-generation, and third-generation super porous hydrogels (SPH). Acrylate-based SAPs and their derivatives are the first-generation SPH, whereas the super porous hydrogel

R. Pattanayak (✉) · T. Jena
CIPET: SARP-LARPM, Bhubaneswar 751024, India
e-mail: roshnipattanayak@gmail.com

S. Pradhan and S. Mohanty (eds.), *Bio-based Superabsorbents*, Engineering Materials,
https://doi.org/10.1007/978-981-99-3094-4_10

composite with crosslinked composite agents is the second-generation SPH. The third generation SPH can be developed by adding the hybrid agent to the SPH to form an interpenetrating polymeric network to form a three-dimensional crosslinked structure. In 1938, the first water-absorbent polymer was synthesised by W. Kern by using acrylic acid with divinylbenzene in an aqueous medium. Later, hydroxy alkyl methacrylate-based hydrogel material was prepared for the contact lens application, and potassium acrylate crosslinked material was synthesised for the development of a water immobilising agent useful in fire-fighting application. The first commercialised SAPs, starch graft-Polyacrylonitrile, was produced in the 1970s at the Northern Regional Research Laboratory, US. After that, Japan, France and Germany commercialised the superabsorbent polymer for use in feminine napkins and baby diaper applications [1, 2]. In these products, mainly acrylate-based synthetic superabsorbents were substituted in bulky cellulosic cotton to absorb a large amount of body fluid. In this way, SAP leads to a significant revolution in the personal healthcare industry. Apart from this, these can be useful in agriculture, the food packaging industry, coating material, the waste management industry, medical devices and tissue engineering applications. However, the use of more amount synthetic petrochemical-based superabsorbents in various applications brings an adverse effect on the environment as well as human health. According to a recent report, about 12.3 billion menstrual pads are being thrown into landfill in a single year. The acrylate-based SAPs used in menstrual napkins takes 500–700 years to completely decompose in landfill which brings a great matter of concern among scientist and their interest goes to the bio-based superabsorbent to a vial alternative for overcoming this problem. These bio-based SAPs are of two types Polysaccharide based and protein-based superabsorbent polymers. Many of the ongoing studies focus on the development of superabsorbent polymers from natural sources and using them in a variety of applications. In this context, we have provided a brief overview of the current trends in bio-based SAPs and their potential future aspects in various ranges of applications. This is schematically represented in Fig. 1.

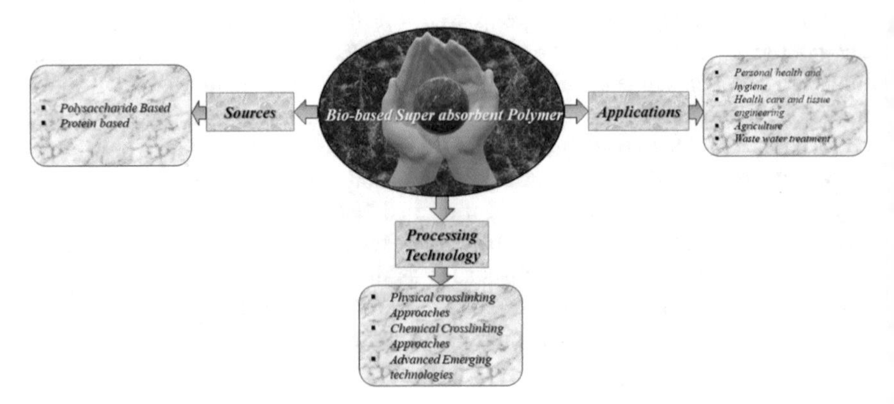

Fig. 1 Bio-based superabsorbent polymer

2 Sources of Bio-based SAP

Bio-based SAPs are sourced from living organisms; abundantly in plants and animals and comprise two categories such as (a) Polysaccharides-Based SAPs and (b) Protein-Based SAPs. These superabsorbents are natural source absorbent polymers and their utilisation has sparked a lot of scientific interest due to their biocompatible and biodegradability properties. Apart from this, these are non-toxic and easily available as raw materials and the processability cost is low which makes them ideal over synthetic ones. A comparison between bio-based and petroleum-based SAPs is represented in Fig. 2. Recently, most of the developed SAPs are being synthesised bio-based to adopt green sources which would be potential eco-friendly alternatives to the environment.

2.1 *Polysaccharide-Based SAPs*

Monosaccharides macromolecules aggregately form heterogeneous amorphous water-insoluble polysaccharide polymers which are significantly elaborated forms of carbohydrates in the living organism. These include cellulose, starch, chitosan, carrageenan and alginates which are cheaper, biocompatible, biodegradable and renewable materials with greater absorbency. Because of these properties, they become a prominent alternative over petroleum-based synthetic SAPs widely utilised in various applications including agriculture, building materials, cosmetics, medical products, and personal hygiene products.

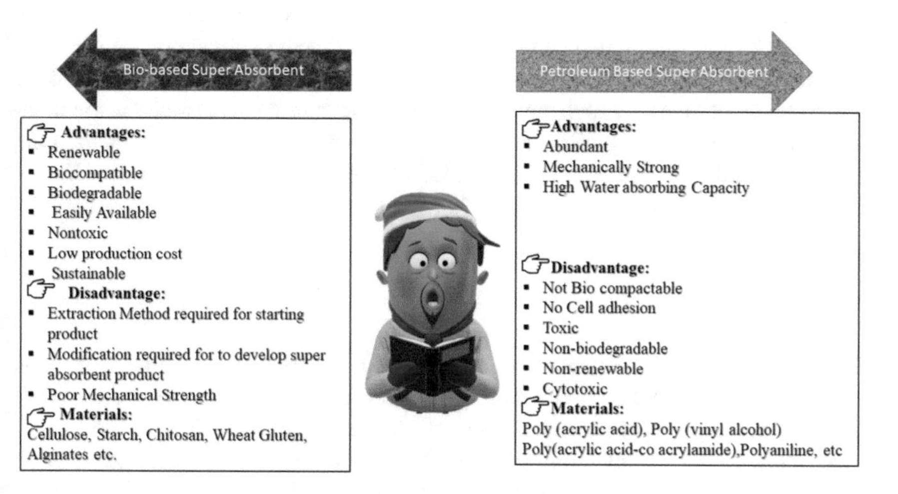

Fig. 2 Comparison of bio-based SAPs and petroleum-based SAPs

2.1.1 Cellulose

Cellulose ($C_6H_{10}O_5$) is a natural organic polymer sourced from the living organism; abundantly from plant sources consisting of β-linkage D-glucose units with a large amount of hydroxyl group (–OH group) which contributes to a significant absorbency to the product. The bio-based cellulose is mainly sourced from plant cellulose and some from bacterial cellulose which is synthesised from *Acetobacter xylinum* bacteria. Hydroxyethyl cellulose, Hydroxymethyl cellulose, Hydroxypropyl Methyl Cellulose and Carboxymethyl cellulose are the most widely used cellulosic materials in the articulation of bio-based SAPs. The presence of cellulose in their backbone influences the swelling behaviour in absorbing water and other aqueous fluids, creating the three-dimensional network hydrogel structure without being insoluble in them. The excellent biocompatible and biodegradable properties of this renewable source cellulose have encouraged its use in the development of bio-based products. These are utilised in agricultural products, building materials, skin care products, the paper industry, diapers, tampons and sanitary napkin products. Firstly, in the nineteenth century, cotton-based personal healthcare products were developed wrapped in non-woven rayon cover sheets. Eventually, a huge revolution was visualised in the technology and novel tactic incorporation of cellulose derivatives hydrogel is used as SAP particles for various fields of applications. Liu et al. [3] developed SAP particles from the waste of novel flax yarn which showed a greater biodegradability property with an excellent blood absorbency as compared to commercially available products. Another novel cellulose-based hydrogel was developed by Ahmad Mahmoodzadeh et al. and modified with silica aerogel and calcium chloride. This hydrogel superabsorbent acts as an effective engineered material which shows a significant biocompatible haemostatic material suitable for controlling the blood from the wound. This showed 60 g/g blood absorbency with suitable biocompatible properties which would be a better option for the clinical approach [4]. Apart from this, novel cellulose-based SAPs can be a potential biomimetic material utilised for the treatment of obesity which manifests a higher magnitude of elasticity to the functional fibres and almost similar elasticity to the vegetables beneficial in maintaining gut tissue, and this first product was aided as weight management in the US and Europe [5].

2.1.2 Starch

Starch is a polymeric polysaccharide comprised of several glucose units such as amylose and amylopectin linked with glycosidic bonds preferably found in plants acting as an energy source in it. It is economically feasible, renewable and biodegradable, and the presence of a higher amount of hydroxyl groups in the composition of starch leads to a large amount of swelling, thereby making it a good adsorbent. Thus, bio-based starch superabsorbent can be utilised in various fields such as agriculture, personal health care application, drug delivery, tissue engineering application, construction building component and wastewater purification applications. A published report by Kang Zhong revealed that Sulfonated corn starch/poly (acrylic acid)-based superabsorbent showed 498 and 65 g g^{-1} of absorbency through distilled water and saline solution utilising as a potential composition for plant fertiliser for its excellent prolonged nutrient absorbency. Grafting in Starch enhances the absorbency and Starch grafted with polyacrylamide showed the highest amount of absorbency but was restricted in implementation to the application due to its lower mechanical gel strength and higher cost of production. Jihuai Wu et al. developed a superabsorbent composite of starch-grafted polyacrylamide with aluminium silicate clay to increase its absorbency to about 3000 gm of water absorbency in just 30 min with excellent compressive properties [6]. Another novel starch/itaconic acid superabsorbent (63% bio-based content) showed 650 g/g of water absorbency and good biodegradable property which could be utilised as a good greener alternative to petroleum-based hydrogels [7].

2.1.3 Chitosan/Chitin

Chitosan is the most commonly used biopolymer from the polysaccharide group derived from chitin by the deacetylation process. These are mainly structural elements profoundly extracted from the exoskeleton of crabs and shrank. The presence of the hydroxyl group (–OH), as well as the amino group ($-NH_2$), is responsible for the higher amount of water absorption. Apart from this, chitosan shows good antimicrobial, biocompatible, biodegradable and non-toxic properties for which it becomes a potential candidate for various applications such as biosorbent for wastewater treatment, agriculture, personal hygiene and tissue engineering health care applications. The swelling capacity of the chitosan-based material is higher for cationic behaviour in crosslinked structure molecules. Jie Cao et al. developed novel chitosan-based porous gel beads which showed absorbency of 3675 g/g in water and 170 g/g in a 0.15 mol/L NaCl solution [8]. Another two consecutive studies performed by Narayanan et al. showed that novel chitosan-based SAPs, namely CHCAUR (chitosan, citric acid and urea) and CHEDUR (chitosan, EDTA and urea) showed 1250 and 500 g/g of water absorbency respectively which will be useful for agricultural applications. In addition to this, they also currently developed a new combination of chitosan superabsorbent using itaconic acid and succinic acid individually which provide a good result of absorbency [9].

2.1.4 Alginate

Alginic Acid is a natural acidic polysaccharide biopolymer that is generally chemically extracted from brown algae and its composition consists of two monomeric monosaccharides such as (1,4)-α-L-guluronic acid (G) and (1,4)-β-D-mannuronic acid (M). This bio-source has a wide range of applications in textile, dental, medical health care and wound dressing applications. Many researches have been conducted on sodium alginate-based composites because of their significant swelling and acidic property. Yizhe Wang et al. reported a novel alginate-based composite that showed 532 g/g of swelling capacity with a good porous structure [10].

2.2 Protein-Based SAPs

Natural SAPs derived from protein polymer can act as a significant substitute for synthetic SAPs. Because of their high polymerization capacity and versatile primary, secondary, and tertiary structures, they can be easily altered to introduce enhancements which make proteins a viable alternative to synthetic SAPs. Keratin, gelatin, wheat gluten and collagen are some of the widely researched proteins aimed by researchers to develop bio-based SAPs. These bio-based alternatives have proved to be superior alternatives for synthetic SAPs as well as replaced the environmental concerns brought by acrylamide content in previously used synthetic SAPs. The biosafety of the material allows them to be useful in the fields of agriculture, biomedical engineering and pollution control.

Reports have shown some proteins displaying a significant rise in their absorbency capacity. The maximum swelling ratio was recorded by keratin protein at pH 9, at the end of 48 has 1791% [11]. Another keratin-based superabsorbent resin shows the highest water absorbency in distilled water and 0.9 wt% and NaCl solutions in 714.22 and 70.08 g/g capacity, respectively [12]. Collagen showed a maximum capacity of swelling in distilled water under the optimised conditions, which was found to be 268 g/g [13]. A gelatin blend hydrogel showed an absorption capacity of 375 g/g, the results of which were established with Design-Expert. The nanocomposite was shown to have better biodegradability compared to that of the copolymer hydrogel [14]. For the gelatin-based hydrogel, the one with the highest acrylamide content displayed a swelling capacity of 115% [15].

3 Processing Technology from Development to Implementation

The negative impact of polyacrylate-based hydrogel on the health and environment brings attention towards the bio-based superabsorbent. In the twenty-first century, the production of these bio-based SAP has increased in bulk amount because of its substantial demands from agriculture to industry level. Many technologies have been adopted by researchers to achieve the desired superabsorbent material. In these technological approaches, ionic bonds, covalent bonds or weak and strong physical attraction plays an important aspect to form the desired product. These methods are

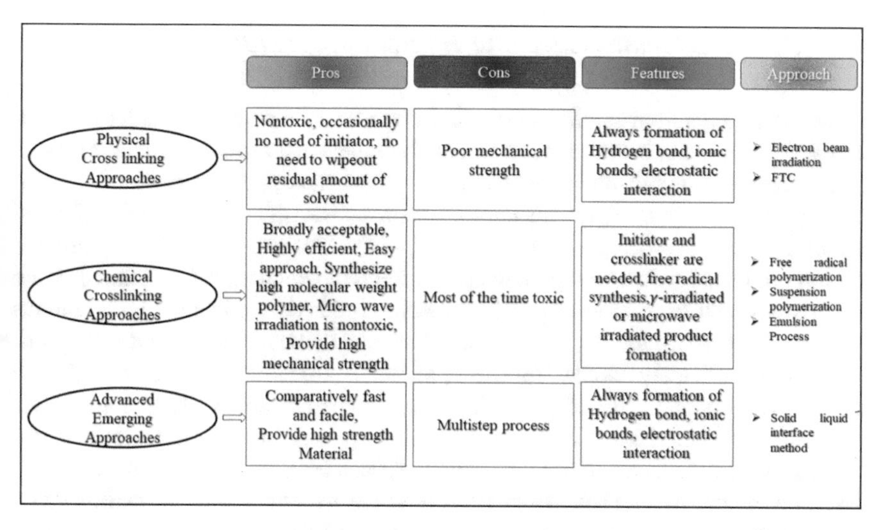

Fig. 3 Various methods for synthesis of SAPs and their pros and cons

divided into two categories as physical crosslinking approach (PCA) and the chemical crosslinking approach (CCA). PCA includes electron beam irradiation, freeze-thawing and hydrogen bond crosslink, whereas CCA includes suspension polymerization and free radical polymerization. Apart from these, many other advanced emerging approaches such as in-situ photopolymerization, solid–liquid interface technique, etc. are also considered for the development of superabsorbent material. Different methods for the synthesis of SAPs and their pros and cons are described in Fig. 3.

Besides these, bio-based membranes, thin films or sheets are being prepared by using natural superabsorbents which can able to absorb the higher amount of liquids applicable in various absorbent applications. For this, solvent casting, conventional phase inversion or electrospinning techniques are used which is reliable, cost-effective and easy to fabricate the bio-based superabsorbent material.

4 Application of Bio-based SAP

The bio-based superabsorbent polymers are having following applications.

4.1 Personal Health Care and Hygiene Applications

Earlier, first-generation acrylate-based SAPs were used in personal health care products such as diapers, incontinence pads, adult diapers and sanitary napkin products. Due to their low degradability, recyclability, low efficiency and toxic behaviour, bio-based products are now a growing choice of consumers which replace the traditional absorbing materials such as cotton, paper, cloth, etc. The researchers are investing in natural products for developing biodegradable and biocompatible SAPs and they are majorly sourced from biowaste and processing industries indicating a continuous stream of supplying for raw materials. Nowadays cellulose-based hydrogel products have been widely adopted for this application. According to a recent article, a novel environmentally friendly cellulose-based superabsorbent was developed via free radical polymerization which showed an absorbency of 475 g/g with superior biodegradability that can be used in agriculture as a favourable alternative to fossil-based petroleum SAPs [16]. Apart from this, the nanofiber form of carboxylated wheat straw Cellulose shows significant mechanical strength and a good swelling capacity of about 200 g/g amount applicable to hygiene applications.

4.2 Cosmetics and Skincare Products

In general, SAP possesses higher water content, which can be used in cosmetics products for its excellent moisturising properties. The bio-based hydrogel SAP provides an active hydrophilic matrix to the skin providing an active ingredient for longer times on the applicated site with maintaining a high localised active agent concentration in the surrounding tissues. Bioadhesive hydrogel derived from natural polymer sources like xanthan gum, guar gum and carrageenan provides an active ingredient for protecting the tissues of the skin very nicely. A newly developed polysaccharide-based hydrogel product derived from xanthan gum, pullulan, carrageenan and konjac mannan showed a significant flexible transparent and adherent property which could be useful as a better skin care product [17].

4.3 Drug Delivery System

The ongoing research in the present day focuses on finding enhanced technologies to achieve more efficient drug delivery throughout the desired body site. A variety of Super absorbent-based biomaterials are now worked extensively on to develop drug carriers that can provide controlled drug release at the required site, before facing degradation. According to a study, collagen has shown efficient drug release in the tumour microenvironment based on activation with nitric oxide (NO), and this

NO activates the matrix metalloproteinase which helps in the controlled degradation of collagen [18]. Collagen-grafted polyacrylic acid copolymerise with methyl acrylic acid hydrogel has shown controlled release of orally administered drugs in the human intestine in acidic conditions, suitable for the proteolytic environment in the gastrointestinal region [19]. Gelatin nanoparticles responsive to internal stimuli are targeted for cancer drug delivery [20] and silk fibroin has also shown remarkable competence in effective drug delivery [21].

4.4 Agriculture

A great deal of advancement in incorporating fertiliser sustainability in agriculture has come forward due to the development of an SAP-based delivery system. Many studies have tried to develop bio-based SAP polymers that can efficiently deliver nutrients and fertilisers to the crops for their easy uptake. Another major point of focus is improving the water retention capacity of the soil by adding SAP particles. To avoid any kind of undesirable effect on the quality of soil or crops growing on it, biomaterials are mainly being deployed to develop these SAPs. PLA/cellulose composite [22], starch-based hydrogel reinforced with char particles [23] and chitosan-derived composite materials [24] are among the many others which have shown remarkable water absorption under load and different pH variations, implying the potential for agricultural applications.

4.5 Wound Dressing

Wound dressing mainly targets covering a wound/lesion to prevent it from getting in contact with any foreign particles and keeping the site clean. A natural inclination towards using organic material for this purpose has led to increasing research in developing wound dressing material from plant and animal sources. Bacterial cellulose has been deemed one of the leading candidates for developing material for wound dressing. The material can be easily manipulated to give variation in its physical properties such as porosity and tensile strength to obtain optimum characteristics similar to the tissue being replaced [18]. Other biomaterials being used for similar purposes include silk fibroin and gelatin. While fibroin provides an ideal substratum for the growth, differentiation and adhesion of skin cells, gelatin serves as an ideal support for cell regeneration due to its mechanical, physiochemical and thermal properties [19]. Chitosan-based dressing materials have shown high biocompatibility and have a greater potential to stop blood loss and initiate blood clotting in about 15 s of exposure [25]. Also, keratin-infused silver nanoparticles has proved to have accelerated the wound healing process when used as dressing material by arranging Wound inflammatory response based on macrophages [26].

4.6 Waste Water Treatment

Removing toxic pollutants from water effluent has been an important use of SAPs. Major pollutants of water bodies include heavy metal ions and dyes. One of the chief studies carried out in wastewater treatment is the removal of CV from effluent, an extremely toxic compound major used in the dyeing industry. SAP hydrogel produced from the combination of acrylamides and biomaterial such as sodium alginate using a crosslinker and initiator showed remarkable CV absorption, swelling capacity, easy synthesis and high recycling capacity [27] Straw cellulose-based hydrogel was found effective in increasing water retention capacity of the soil by 681.3 g/g and also demonstrated enhanced release rates of nitrogen and phosphorous in the soil [28].

4.7 Tissue Engineering Application

Biomaterials have been the choice of target for being used to make scaffolds, required for tissue engineering for their superior biocompatible nature. Due to superior water absorption and life-supporting characteristics, they can bind with cell receptors and facilitate the growth of cell [29] PCL/PLA [30], PVP/PAA [31], collagen [32], gelatin [33] and chitosan [34] are now the focus of research for many scientists to develop biocompatible scaffolds. Studies on these materials have shown them to exhibit superior tensile and porosity characteristics and also show impressive cell viability. They are being used extensively for bone and cartilage repair, muscle tissue repair, nerve tissue regeneration and vascular grafts.

5 Future Scope

Bio-based SAPs are having non-toxic, cheap costs, significant hydrophilicity and biodegradable properties with biocompatible and limited adverse effects on the environment. Thus, these would be a greener alternative to the recognised hazards associated with petroleum-based synthetic SAPs. Researchers are currently concentrating on the rise in popularity of environmentally friendly products, giving bio-based SAPs more attention, in the hopes that in the near future, bio-based materials will completely replace petroleum-based SAPs in the world of superabsorbents. According to a published report in the Archives of environmental health by the Environmental Health Association of Nova Scotia states that toxic volatile organic chemical compounds (VOC) such as xylene, toluene, benzene and dipentene released from diaper products create an adverse effect on human health upon long term exposure to the skin triggering allergic reactions and inflammations which are somehow cancerous to the humans. Nowadays the development of bio-based SAPs is still a small initiative fraction of the ongoing global SAP demands in the market. But to emphasise the attention to a large amount of waste-affected environmental issues and health-related problems, bio-based superabsorbents are the real potential solution to overcoming these issues. However, despite all the merits of green-sourced products, these are still having some drawbacks which would be a significant challenge that needs special attention for commercialization of it in larger amounts.

6 Conclusion

SAPs have proved their practicality in a wide range of applications, owing to their characteristic water-holding capacity. An immediate need to develop environment-friendly SAPs has risen due to their increased use. Currently developed biomaterial-based composites/copolymers are being further processed by different polymerisation techniques and synthesis methods and studies are ongoing to provide substantial mechanical strength to them to make them a better substitute for the required applications. Our future aim will be to head towards more efficient biomaterials, eventually progressing towards next-generation smart biomaterials which will have the capacity to sense and respond to their environment as well. These materials would be donned with better biocompatibility as well as innovative optical and chemical properties, keeping in eye the needs of tissue engineering. Biomaterials made with the purpose of pollution control can come up with ways to enhance the ability to detect pollutants as well as increase pollutant uptake capacity. Similarly, in the field of agriculture, better control over the release of fertilising compounds would make SAPs more reliable.

References

1. Venkatachalam, D., Kaliappa, S.: Superabsorbent polymers: a state-of-art review on their classification, synthesis, physicochemical properties, and applications. Rev. Chem. Eng. (2021). https://doi.org/10.1515/REVCE-2020-0102
2. Behera, S., Mahanwar, P.A.: Superabsorbent polymers in agriculture and other applications: a review. Polym. Plast. Technol. Mater. **59**(4), 341–356 (2020). https://doi.org/10.1080/25740881.2019.1647239
3. Pittler, M.H., Ernst, E.: Dietary supplements for body-weight reduction: a systematic review. Am. J. Clin. Nutr. **79**(4), 529–536 (2004). https://doi.org/10.1093/AJCN/79.4.529
4. Mahmoodzadeh, A., Moghaddas, J., Jarolmasjed, S., Ebrahimi Kalan, A., Edalati, M., Salehi, R.: Biodegradable cellulose-based superabsorbent as potent hemostatic agent. Chem. Eng. J. **418**, 129252 (2021). https://doi.org/10.1016/J.CEJ.2021.129252
5. Madaghiele, M., Demitri, C., Surano, I., Silvestri, A., Vitale, M., Panteca, E., Zohar, Y., Rescigno, M., Sannino, A.: Biomimetic cellulose-based superabsorbent hydrogels for treating obesity. nature.com (2021) [Online]. https://www.nature.com/articles/s41598-021-00884-5. Accessed 13 Dec 2022

6. Wu: Synthesis and properties of starch-graft-polyacrylamide/clay superabsorbent composite. Macromol. Rapid Commun. (Wiley Online Library) (2000). https://onlinelibrary.wiley.com/doi/abs/10.1002/1521-3927(20001001)21:15%3C1032::AID-MARC1032%3E3.0.CO;2-N. Accessed 13 Dec 2022
7. Bora, A., Karak, N.: Starch and itaconic acid-based superabsorbent hydrogels for agricultural application. Eur. Polym. J. (Elsevier) (2022) [Online]. https://www.sciencedirect.com/science/article/pii/S0014305722004347. Accessed 13 Dec 2022
8. Cao, J., Tan, Y., Che, Y., Ma, Q.: Fabrication and properties of superabsorbent complex gel beads composed of hydrolyzed polyacrylamide and chitosan. J. Appl. Polym. Sci. **116**(6), 3338–3345 (2010). https://doi.org/10.1002/APP.31796
9. Sangeetha, E., Narayanan, A., Dhamodharan, R.: Super water-absorbing hydrogel based on chitosan, itaconic acid and urea: preparation, characterization and reversible water absorption. Polym. Bull. **79**(5), 3013–3030 (2022). https://doi.org/10.1007/S00289-021-03641-W
10. Wang, Y., Wang, W., Shi, X., Wang, A.: Enhanced swelling and responsive properties of an alginate-based superabsorbent hydrogel by sodium p-styrenesulfonate and attapulgite nanorods. Polym. Bull. **70**(4), 1181–1193 (2013). https://doi.org/10.1007/S00289-012-0901-0
11. Arican, F., Uzuner-Demir, A., Sancakli, A., Ismar, E.: Synthesis and characterization of superabsorbent hydrogels from waste bovine hair via keratin hydrolysate graft with acrylic acid (AA) and acrylamide (AAm). Chem. Pap. **75**(12), 6601–6610 (2021). https://doi.org/10.1007/S11696-021-01828-Z
12. Sci-Hub | Synthesis and swelling behaviors of semi-IPNs superabsorbent resin based on chicken feather protein. J. Appl. Polym. Sci. **131**(1), n/a–n/a. https://doi.org/10.1002/app.39748. https://sci-hub.se/10.1002/app.39748. Accessed 28 Oct 2022
13. Sadeghi, M., Hosseinzadeh, H.: Synthesis and super-swelling behavior of a novel low salt-sensitive protein-based superabsorbent hydrogel: collagen-g-poly (AMPS). Turk. J. Chem. (journals.tubitak.gov.tr) **34**(5), 739–752 (2010). https://doi.org/10.3906/kim-0910-21
14. Nath, J., Chowdhury, A., Ali, I., Dolui, S.K.: Development of a gelatin-g-poly(acrylic acid-co-acrylamide)–montmorillonite superabsorbent hydrogels for in vitro controlled release of vitamin B12. J. Appl. Polym. Sci. (Wiley Online Library) **136**(22) (2019). https://doi.org/10.1002/app.47596
15. Serafim, A., et al.: One-pot synthesis of superabsorbent hybrid hydrogels based on methacrylamide gelatin and polyacrylamide. Effortless control of hydrogel properties through. pubs.rsc.org 1–3 (2013). https://doi.org/10.1039/x0xx00000x
16. Arredondo, R., et al.: Performance of a novel, eco-friendly, cellulose-based superabsorbent polymer (cellulo-SAP): absorbency, stability, reusability, and biodegradability. Can. J. Chem. Eng. (2022). https://doi.org/10.1002/CJCE.24601
17. Hydrogel for natural cosmetic purposes (2011)
18. Dong, X., et al.: Enhanced drug delivery by nanoscale integration of a nitric oxide donor to induce tumor collagen depletion. Nano Lett. **19**(2), 997–1008 (2019). https://doi.org/10.1021/ACS.NANOLETT.8B04236
19. Noppakundilograt, S., Choopromkaw, S., Kiatkamjornwong, S.: Hydrolyzed collagen-grafted-poly[(acrylic acid)-co-(methacrylic acid)] hydrogel for drug delivery. J. Appl. Polym. Sci. **135**(1) (2018). https://doi.org/10.1002/APP.45654
20. Hussain, A., Hasan, A., Babadaei, M.M.N., Bloukh, S.H., Edis, Z., Rasti, B., Sharifi, M., Falahati, M.: Application of gelatin nanoconjugates as potential internal stimuli-responsive platforms for cancer drug delivery. J. Mol. Liq. (Elsevier) (2020) [Online]. https://www.sciencedirect.com/science/article/pii/S0167732220309296. Accessed 15 Dec 2022
21. Tomeh, M.A., Hadianamrei, R., Zhao, X.: Silk fibroin as a functional biomaterial for drug and gene delivery. mdpi.com **11**(10) (2019). https://doi.org/10.3390/pharmaceutics11100494
22. Calcagnile, P., Sibillano, T., Giannini, C., Sannino, A., Demitri, C.: Biodegradable poly(lactic acid)/cellulose-based superabsorbent hydrogel composite material as water and fertilizer reservoir in agricultural applications. J. Appl. Polym. Sci. **136**(21) (2019). https://doi.org/10.1002/APP.47546

23. Motamedi, E., Motesharezedeh, B., Shirinfekr, A., Samar, S.M.: Synthesis and swelling behavior of environmentally friendly starch-based superabsorbent hydrogels reinforced with natural char nano/micro particles. J. Environ. Chem. Eng. (Elsevier) (2020) [Online]. https://www.sciencedirect.com/science/article/pii/S2213343719307067. Accessed 15 Dec 2022
24. Said, M., Atassi, Y., Tally, M., Khatib, H.: Environmentally friendly chitosan-g-poly(acrylic acid-co-acrylamide)/ground basalt superabsorbent composite for agricultural applications. J. Polym. Environ. **26**(9), 3937–3948 (2018). https://doi.org/10.1007/S10924-018-1269-5
25. Akram, A.M., Omar, R.A., Ashfaq, M.: Chitosan/calcium phosphate-nanoflakes-based biomaterial: a potential hemostatic wound dressing material. Polym. Bull. (Springer) (2022). https://doi.org/10.1007/s00289-022-04300-4
26. Konop, M., et al.: Evaluation of keratin biomaterial containing silver nanoparticles as a potential wound dressing in full-thickness skin wound model in diabetic mice. J. Tissue Eng. Regen. Med. **14**(2), 334–346 (2020). https://doi.org/10.1002/TERM.2998
27. Rehman, T., et al.: Fabrication of stable superabsorbent hydrogels for successful removal of crystal violet from waste water. pubs.rsc.org [Online]. https://pubs.rsc.org/en/content/articlehtml/2019/ra/c9ra08079a. Accessed 15 Dec 2022
28. Wang, W., Yang, Z., Zhang, A., Yang, S.: Water retention and fertilizer slow release integrated superabsorbent synthesized from millet straw and applied in agriculture. Ind. Crops Prod. (Elsevier) (2021) [Online]. https://www.sciencedirect.com/science/article/pii/S0926669020310438. Accessed 15 Dec 2022
29. Behera, S., Mahanwar, P.A.: Superabsorbent polymers in agriculture and other applications: a review. Polym. Plast. Technol. Mater. (Taylor & Francis) **59**(4), 341–356 (2020). https://doi.org/10.1080/25740881.2019.1647239
30. Wang, L., Wang, D., Zhou, Y., Zhang, Y., Li, Q., Shen, C.: Fabrication of open-porous PCL/PLA tissue engineering scaffolds and the relationship of foaming process, morphology, and mechanical behavior. Polym. Adv. Technol. **30**(10), 2539–2548 (2019). https://doi.org/10.1002/pat.4701
31. Demeter, M., Meltzer, V., Călina, I., Scărișoreanu, A., Micutz, M., Albu Kaya, M.G.: Highly elastic superabsorbent collagen/PVP/PAA/PEO hydrogels crosslinked via e-beam radiation. Radiat. Phys. Chem. **174** (2020). https://doi.org/10.1016/j.radphyschem.2020.108898
32. Dinescu, S., Albu Kaya, M., Chitoiu, L., Ignat, S., Kaya, D.A., Costache, M.: Collagen-Based Hydrogels and Their Applications for Tissue Engineering and Regenerative Medicine, pp. 1–21 (2018). https://doi.org/10.1007/978-3-319-76573-0_54-1
33. Zhao, X., et al.: Photocrosslinkable gelatin hydrogel for epidermal tissue engineering. Adv. Healthc. Mater. **5**(1), 108–118 (2016). https://doi.org/10.1002/ADHM.201500005
34. Nezhad-Mokhtari, P., Akrami-Hasan-Kohal, M., Ghorbani, M.: An injectable chitosan-based hydrogel scaffold containing gold nanoparticles for tissue engineering applications Int. J. Biol. Macromol. **154**, 198–205 (2020). https://doi.org/10.1016/j.ijbiomac.2020.03.112

Correction to: Smart Superabsorbents and Other Bio-based Superabsorbents

Shubhasmita Rout

Correction to:
Chapter 8 in: S. Pradhan and S. Mohanty (eds.), *Bio-based Superabsorbents*, Engineering Materials, https://doi.org/10.1007/978-981-99-3094-4_8

The book was inadvertently published with an incorrect affiliation in chapter 8. The affiliation has been updated as CIPET-IPT, Bhubaneswar, Odisha, India. The correction chapter and the book have been updated with the changes.

The updated version of this chapter can be found at
https://doi.org/10.1007/978-981-99-3094-4_8

S. Pradhan and S. Mohanty (eds.), *Bio-based Superabsorbents*, Engineering Materials,
https://doi.org/10.1007/978-981-99-3094-4_11

Printed in the United States
by Baker & Taylor Publisher Services